HUMAN OSTEOLOGY:
A Laboratory and Field Manual
of the
Human Skeleton

by

WILLIAM M. BASS, PH.D.
Professor of Physical Anthropology and
Head, Department of Anthropology
University of Tennessee
Knoxville, Tennessee

1971

Special Publications
David R. Evans, Editor
Missouri Archaeological Society
15 Switzler Hall
University of Missouri
Columbia, Missouri

Library of Congress No. 77-172091
Second Printing 1973.

In appreciation to my Doctoral Committee

From The University of Pennsylvania:

Carlton S. Coon
Loren C. Eiseley
Wilton M. Krogman (Chairman)

From The Smithsonian Institution:

Frank H. H. Roberts, Jr.
T. Dale Stewart
Waldo R. Wedel

PREFACE

I have long been interested in the human skeleton and for some years have taught physical anthropology and anatomy. In helping students to understand anatomical form and structure, landmarks and terminology, I have become increasingly aware that although physical anthropology is one of the oldest studies of man, there are few if any publications directed specifically to workers in this field that give sufficient information for the identification and analysis of the bones of the human skeleton.

The present book is an attempt to remedy this deficiency. It does not pretend to be a complete summary, for such would take much more space than is desirable in a manual. It does attempt to present basic anatomy of the bones, major anatomical landmarks, criteria for determining right or left sides of paired bones, basic anthropometric measurements and indices and comparative data drawn from the literature to make the indices, and thus the anthropometric analysis, more meaningful. Variations or anomalies are given for a few bones along with a short list of references to the literature for most bones.

This manual is intended for the use of both students and professionals in the field as an aid to the identification of human bone and in the laboratory as an aid to identification, study and analysis. The arrangement is such that it can be used as a text in the osteological section of either an introductory or advanced course in physical anthropology or archaeology.

I acknowledge with pleasure the assistance of my many students who, over the past eleven years, have offered criticisms and suggestions on how to present the data in a clear and understandable manner. Three books have been published by colleagues in this general area of science but none with the same approach as this manual: J. E. Anderson - "The Human Skeleton, Manual for Archaeologists"; Jack Kelso and George Ewing - "Introduction to Physical Anthropology Laboratory Manual"; and D. R. Brothwell - "Digging Up Bones." To them I acknowledge some ideas regarding format. None approach the identification of the human skeleton from the aspect of growth as is done in this manual.

Usually about half of the human skeletons from an archaeological site are subadults. Without this manual it is difficult to find illustrations and information to identify subadult material. In addition, this manual presents each bone separately, along with information on its growth, age, sex and measurements. This makes the manual easier to use.

To Dr. T. Dale Stewart I am indebted for reading the manuscript and making many meaningful suggestions as to format and content, and for his encouragement and help over the years. Dr. M. Dale Kinkade, Dr. William Clemens and Mr. Douglas H. Ubelaker have read the manuscript and offered many helpful suggestions. In addition Mr. Ubelaker prepared most of Chapter IV on the Human Dentition. I appreciate the work of Mrs. Madelyn Jenks in the preparation of the manuscript. Miss Peggy Grinvalsky illustrated most of the bones.

CONTENTS

INTRODUCTION

WHY STUDY BONES?

Often the question is asked "Why study bones?" A few of the more obvious of the many reasons are listed below.

1. They constitute the evidence for the study of fossil man.
2. They are the basis of racial classification in prehistory.
3. They are the means of biological comparison of prehistoric peoples with the present living descendants.
4. They bear witness to burial patterns and thus give evidence of culture and world view of the people studied.
5. They form the major source of information on ancient diseases and often give clues as to the causes of death.
6. Their identification often helps solve forensic cases.

Bones are the framework of the vertebrate body and thus contain much information about man's adaptive mechanisms to his environment. The study of evolution would be almost impossible if bones were eliminated as a source of data. In summary, the answer is that bones often survive the process of decay, and therefore provide the main evidence of the human form after death in time past, and of the customs and diseases that affected them.

BASIC TERMS AND ORIENTATION OF BODY

The exact meaning of certain fundamental terms used in anatomy must be understood and kept in mind (see Fig. 1; note that the terms can apply to both two-legged and four-legged animals). The terms given here are only the most basic; a more comprehensive list is given in Appendix 1.

A. Anatomical Position or Standard Erect Position: standing with the feet forward and the hands at the sides with the palms forward. In this position no long bones are crossed. When the back of the hand is turned forward the radius and ulna in the lower arm are crossed, with the radius being the outside (lateral) bone at the elbow, but the inside (medial) bone at the wrist.

Always orient the bone you are studying as it is in your body.

When you are determining the side from which a bone came, the bone will always be from the same side as it is in your body. With the body in the Standard Erect Position there are three fundamental planes (sections). (Fig. 1)

1. Sagittal or midsagittal—that which divides the body into right and left halves.
2. Frontal or coronal—that which divides the body into front and rear portions.
3. Transverse or horizontal—any place at right angles to 1 and 2.

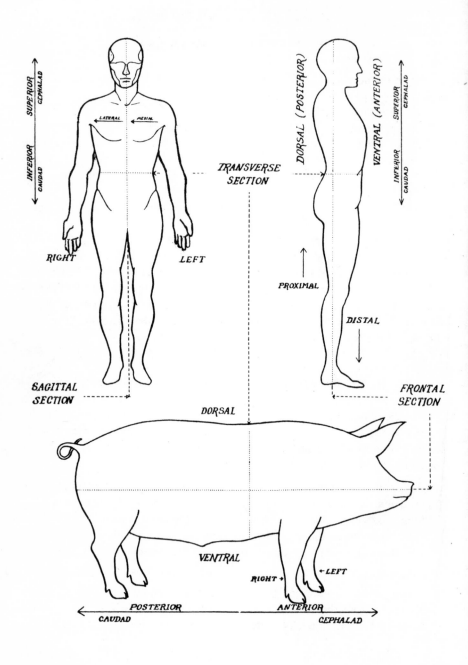

Fig. 1. Basic orientation of the body.

B. Principal directions for parts of:
1. Body
Front—ventral or anterior
Rear—dorsal or posterior
Upper—cranial or superior
Lower—caudal or inferior
Medial—toward the midline
Lateral—away from the midline
2. Limbs
Proximal - that portion or end nearest the trunk or head
Distal - that portion or end farthest from the trunk or head
C. Skeleton (the solid framework of the body) is composed of:
1. Bone. (Ligaments unite the bones in the living.) When bones come together they are said to *articulate.*
2. Cartilage. The tissue covering the surfaces where bones articulate.
D. Four classes of bones:
1. Long bones - main components of limbs; in part sustain weight, and with muscles attached to them form a system of levers for movement.
2. Short bones - metacarpals of hands, metatarsals of feet, and phalanges of hands and feet; found where compactness, elasticity and limited motion are required.
3. Flat bones - cranial bones, innominates and shoulder blade; offer protection and provide wide areas for muscle attachment.
4. Irregular bones - vertebra, carpal (hand) and tarsal (foot) bones and many of the cranial bones; often very complex and with peculiar forms for the functions they perform.

Total number of bones in the adult is usually stated to be 206;(Fig. 2) however, variations may occur as noted in the following (Table 1) of "Bones of the Skeleton".

There are usually 806 centers of ossification in the growing skeleton, but by maturity these centers have united to form the 206 bones of the adult. The surface contour of bone presents irregularities in the shape of eminences and depressions such as:

		Example	
Name	*Description*	*Bone*	*Anatomical Area*
Crest—	a ridge, especially one surmounting a bone or its border	Innominate—iliac crest Radius—interosseous crest	
Spine—	a sharp prominence or slender process of bone	Innominate—anterior superior iliac spine	
Process—	a slender projecting point	Vertebra—spinous process	
Tubercle—	a small tuberosity	Skull—external occipital protuberance	
Groove—	a shallow linear depression	Innominate—obturator groove	

TABLE 1. BONES OF THE ADULT SKELETON

CRANIAL BONES

Single

Frontal	1
Occipital	1
Sphenoid	1
Mandible	1
Ethmoid	1
Vomer	1
Hyoid	1

7

Paired

Parietal	2
Temporal	2
Maxilla	2
Nasal	2
Zygomatic (malar or cheek)	2
Lacrimal	2
Palate	2
Inferior nasal concha	2
Malleus (ear)	2
Incus (ear)	2
Stapes (ear)	2

22

POSTCRANIAL BONES

Single

Cervical vertebrae	7	
Thoracic vertebrae	12	- can be 13, see p. 82
Lumbar vertebrae	5	
Sacrum	1	
Coccyx	1	- can have several segments, see p. 86
Sternum	1	- can be 3 or more parts, see p. 91

27

Paired (Upper extremities)

Scapula	2	
Clavicle	2	
Ribs	24	- variable, see p. 106
Humerus	2	
Radius	2	
Ulna	2	
Carpus	16	
Metacarpus	10	
Phalanges	28	

88

TABLE 1. BONES OF THE ADULT SKELETON (Con't.)

Paired (Lower extremities)

Innominate	2
Femur	2
Patella	2
Tibia	2
Fibula	2
Tarsus	14
Metatarsus	10
Phalanges	28
	62
Total	206

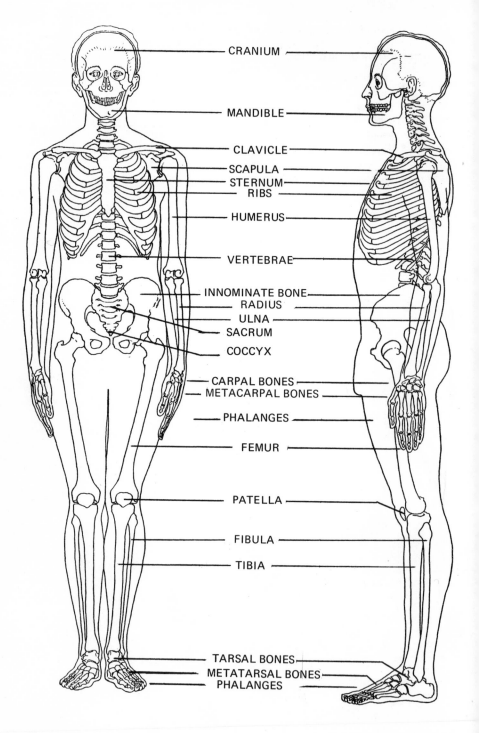

CRANIUM

MANDIBLE

CLAVICLE
SCAPULA
STERNUM
RIBS

HUMERUS

VERTEBRAE

INNOMINATE BONE
RADIUS
ULNA
SACRUM
COCCYX

CARPAL BONES
METACARPAL BONES

PHALANGES

FEMUR

PATELLA

FIBULA

TIBIA

TARSAL BONES
METATARSAL BONES
PHALANGES

Fig. 2.

WHAT BONE IS IT?
STEPS TO FOLLOW IN IDENTIFYING BONES

The student's first encounter with the 206 bones of the human skeleton is often a bewildering experience. What bone is it? How do I start? What should I look for? These are some of the questions asked by the untrained observer. The best procedure is to follow a logical process of elimination of certain bones. The following steps are suggested for identification of bones from the human skeleton. (For a key on how to identify bones from mammalian skeletons see Cornwall 1956: 185-195).

When faced with the task of determining what bone you are looking at, the first step is to decide whether it is subadult or adult. (See Subadult Age Determination, page 13).

Step 1.

Is it cranial or postcranial (literally bones behind the skull in the quadrupedal position)?

Cranial bones: are thin, flat or irregular in shape; if from the vault will have serrated edges (sutures); may contain sinuses (sinuses occur in the frontal, maxilla, zygomatic and sphenoid); are the only bones in which teeth or tooth sockets occur.

If it is a cranial bone the information in Chapter II will aid in the determination of the exact bone.

Step 2.

If it is postcranial, is it a:

a. Long bones:

      6 paired tubular bones, 3 in each limb

          Arm - humerus, radius and ulna

          Leg - femur, tibia and fibula

A long bone has a tubular shaft and an articular area at each end. Compared with other bones of the human body long bones are large (femur is the largest and radius the smallest long bone).

b. Short bones:

      5 metacarpals and 14 phalanges in each hand

      5 metatarsals and 14 phalanges in each foot

      2 clavicles

Short bones resemble long bones in that they have tubular shafts and usually articular areas at both ends (a terminal phalanx does not have an articular surface at the distal end).

c. Flat bones:

      2 hip bones (may apply to some cranial bones)

      2 scapulae

      24 ribs

      1 sternum

Flat bones are large in surface area (for muscle attachments) and usually thin. They have irregularly placed articular surfaces.

d. Irregular bones:

>33 bones of the vertebral column (may apply to some cranial bones)
>8 carpal bones
>7 tarsal bones
>2 patellae

Irregular bones have special shapes for the functions they perform but they are usually thick and short in length. The patella is a sesamoid.

Subadult bones: depending on the age of the subadult individual, the various bones can resemble any of the four classes above. For example, to the untrained, the femur of an infant who died at birth or soon after may resemble a short bone or the bone of a small mammal. Great care should be taken when dealing with subadult bones (bones in which the epiphyses have not attached). The information presented with each bone should be consulted by persons interested in additional information.

CARE AND TREATMENT OF BONES
AND OSTEOMETRIC EQUIPMENT

Human skeletal material used in physical anthropology laboratories is generally obtained from medical supply houses, from dissecting rooms or from archaeological sites. Skeletal remains purchased from medical supply houses are expensive; those obtained from dissecting rooms may include parts that have been sectioned and those from archaeological sites may be incomplete or poorly preserved. In any case, the material should be treated with care and respect at all times.

In this connection the words of J. W. Powell, the founder of the Bureau of American Ethnology may be recalled:

> These materials constitute something more than a record of quaint customs and abhorrent rites in which morbid curiosity may revel. In them we find the evidences of traits of character and lines of thought that yet exist and profoundly influence civilization. Passions in the highest culture deemed most sacred—the love of husband and wife, parent and child, and kith and kin, tempering, beautifying, and purifying social life and culminating at death, have their origin far back in the early history of the race and leaven the society of savagery and civilization alike. At either end of the line bereavement by death tears the heart and mortuary customs are symbols of mourning. The mystery which broods over the abbey where lie the bones of king and bishop, gathers over the ossuary where lie the bones of chief and shamin; for the same longing to solve the mysteries of life and death, the same yearning for a future life, the same awe of powers more than human, exist alike in the mind of the savage and the sage. (Powell 1881: XXVI-XXVII)

The material which you will be handling may be studied by many people for years to come and is much like books in a library. Because of the development of new and scientific techniques which produce increased knowledge, new and possibly significant information can be obtained from previously studied skeletal material. Therefore, care should always be exercised in the handling of skeletal material.

The proper way to carry or hold a skull is with both hands. The only opening into which fingers can safely be inserted is the foramen magnum (large opening) in the base of the skull. Other openings such as the eye orbits are composed of fragile bones which can easily be destroyed by improper handling.

The most common instruments used to measure skeletal material are:

A. The sliding caliper (Plate 1a) is used for certain cranial measurements in the facial region or on the mandible. When the reference points for a measurement are relatively close together and the contours of the skull do not interfere with the process of measuring, the sliding caliper is generally the best choice of instrument.

B. The hinge or spreading caliper (Plate 1b) is devised to enable one to measure points on the skull in which a straight line measuring device is not suitable.

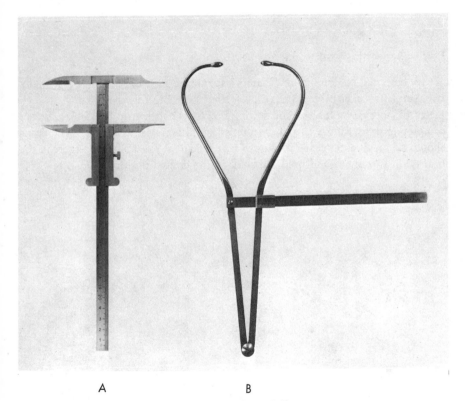

A   B

Plate 1. (A) Sliding Caliper. (B) Hinge (Spreading) Caliper.

C. The osteometric board (Plate 2) is the easiest, most accurate instrument with which to measure the long bones and other large postcranial bones.

These instruments are very expensive because they must be made with great precision. Common sense should dictate the procedure for handling them.

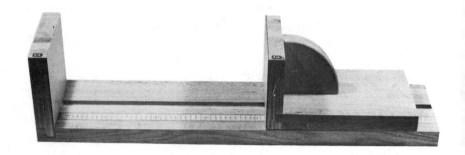

Plate 2. Osteometric Board.

A bean bag or donut ring (Plate 3) should be used to support a skull on a flat surface. A serviceable bag can be made by sewing together two 6 x 6 inch pieces of cloth or by folding and sewing one 6 x 12 inch strip (Plate 3a). Before completing the sewing 1 cup of dry beans should be put into the bag. The beans should be loose enough in the bag so that when the skull is placed thereon the beans will spread out and hold the skull in the desired position.

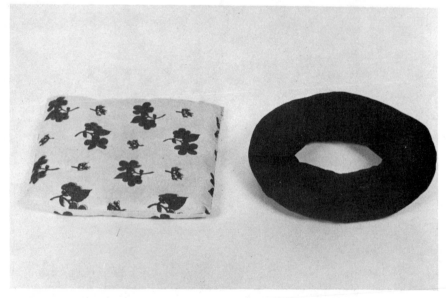

A   B

Plate 3. Bean Bag and Donut Ring.

A donut ring (plate 3b) serves the same purpose as a bean bag but is more difficult to make. It should be approximately 6 inches in diameter with the hole from 2 ½ to 3 inches in diameter. The ring is filled with old rags or stockings to form a soft firm base for the skull.

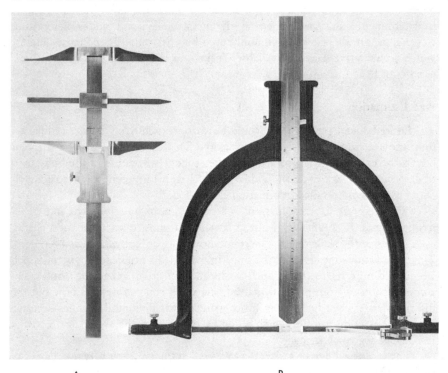

A   B

Plate 4. (A) Coordinate Caliper, (B) Western Reserve Model Head Spanner.

Two anthropometric instruments not as commonly used as A, B, and C above but of great value in taking more difficult measurements are:

D. The Coordinate caliper (Plate 4a) is a sliding caliper with a coordinate attachment for measuring depths below or elevations above two points. The sharp ends of the sliding caliper are placed on two anatomical landmarks and the coordinate attachment is moved so that its sharp point touches the area to be measured. Depths or elevations are read from the millimeter scale on the coordinate attachment.

E. The head spanner (a Western Reserve model designed by Dr. T. Wingate Todd is illustrated in Plate 4b) is designed to take measurements along the mid-sagittal plane with reference to Porion. The ear attachments are placed in both ear holes (external auditory meatus) at Porion and the calibrated measuring rod records distance from the line connecting the two rods. Auricular Height can be taken with the use of the attachment to locate Orbitale thus relating the skull to the Frankfort Horizontal or Plane.

MEASUREMENT OF THE BONES

Many measurements can be taken on the skull and on the long bones, as well as on many of the irregular bones. The length measurements of the long bones may be used to calculate stature, certain other measurements may be used to determine sex and race. In general, the measurements and indices presented in this manual are those most commonly used by physical anthropologists and for which comparative data are available. Following the descriptions of each bone is a list of basic measurements.

Age Estimation

Entire books have been and could be written exclusively on age estimation from the skeleton (for example, Stewart and Trotter 1954). It is not the intent here to be exhaustive, but to acquaint the student with areas of the body used to determine the age of an individual at death and to present a few basic references where more detailed information can be found.

When we say "age estimation" we do not mean how long the individual has been dead. It is difficult, if not impossible, in most cases (without a chemical analysis of the skeleton) to determine how long an individual has been dead. There are various means of estimating time elapsed since death, most of which are derived from related fields, such as stratification from archaeology, and radioactive Carbon-14 from physics. Instead, the main concern here, is how old the individual was when he died; in other words, what biological changes occurred in the skeleton during life which allow us to estimate the age at death with a reasonable degree of accuracy.

The biological age of a skeleton can be determined with varying degrees of success, depending on the period of life reached. At the stage when the teeth are erupting and the epiphyses uniting, age can often be judged quite precisely. After growth has stopped and the permanent dentition has erupted—that is, onwards from about 25 to 30 years—the estimation of age depends almost entirely on degenerative changes.

One of the major factors students should be aware of is human variability. All children crawl before they walk, and walk before they run, but not all children do these at the same time. One child may walk at 9 months, whereas another may not walk until 18 months. This is only to illustrate that people mature at different rates. The same sort of variability exists also in the formation of teeth and bones as Stewart (1963: 264-273) has pointed out.

Historically, we do not find reliable data on age indicators until after the first World War. During the 1920's T. Wingate Todd and others at Western Reserve University began studying such things as epiphyseal union in the documented skeletons from the dissecting rooms. Their object was to establish general biological rules concerning sequential changes in the skeleton during life.

In the 1930's Rudolph Kronfeld (1935) at the Loyola University Dental School summarized the histologic and roentgenographic data on several aspects of tooth formation for both the deciduous and permanent dentition.

By the 1950's anthropologists were becoming increasingly aware of the great range of human variation and the need for more refined techniques of biological aging.

A major problem that exists in obtaining adequate samples of documented skeletons, i.e., skeletons of known sex, race, and age at death. Since documented skeletal populations usually come from anatomical dissecting rooms, they tend to be heavily weighted with individuals in the older age range and from the lower socio-economic levels. A breakthrough of this "age-old" problem, at least for males, was achieved by McKern and Stewart (1957) in their study of skeletal age changes in young Americans killed in the Korean War. Another breakthrough was the studies of Garn and his associates on the variability of tooth formation and its relation to other maturity indicators as seen in successive x-rays taken in life (Lewis and Garn 1960).

Although Krogman had written many short articles on what skeletons have to tell, it was not until 1962 that he brought much of the information together in a textbook entitled "The Human Skeleton in Forensic Medicine" (Krogman 1962).

Within the last few years, techniques for determining skeletal or biological age have become much more refined, and Kerley (1965) has published one of the first microscopic techniques for determining age of human bone. It is my belief that physical anthropology is on the verge of major achievements in the area of skeletal aging. What follows in this manual is only an outline of the developments to date, to acquaint the student with basic techniques. To approach this problem, we should first determine whether the bone(s) on which we are to estimate age is subadult or adult.

Subadult Age Estimation

There are a number of criteria that can be used here:
1. Tooth eruption
2. Epiphyseal closure
3. Length of long bones without epiphyses

Tooth Eruption—The following are *very general* age categories; more specific information should be gained from the following figures (Fig. 3) and from the references.

Birth—usually no teeth erupted

6 mo.—first deciduous teeth usually begin to appear (lower central incisors first)

24 mo.—usually all 20 deciduous teeth have erupted

2-6 yrs.—ossification of the roots of the teeth proceeds, the teeth do not get any larger

As the child grows and the mandible and maxilla get larger, spaces begin to form between the deciduous teeth.

6 yrs.—first permanent teeth, the 6-year molars, appear

6½ yrs.—beginning loss of the deciduous teeth (the central incisors being lost first)

6½-11 yrs.—period of loss of deciduous teeth and replacement by permanent teeth

12 yrs.—second permanent molars (12-year molars) appear

18 yrs.—the third molars (wisdom teeth) may possibly appear. This is a genetically unstable tooth and may never appear, or may erupt beyond 18

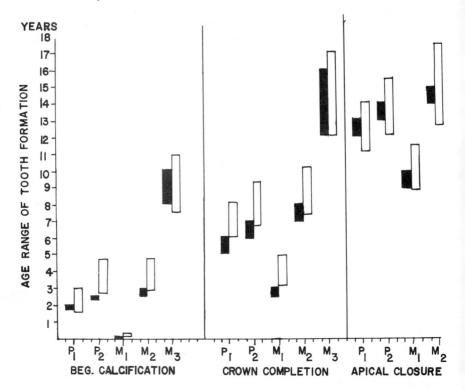

Fig. 3. The variations that exist in tooth formation. The expanding concept of human variability as it relates to tooth formation. Comparison between age ranges of R. Kronfeld (*The Bur*, 35:18, 1935) and 5th-95th percentile limits of S. M. Garn, A. B. Lewis, and D. L. Polacheck (*J. dent. Res.*, 38:135, 1959) for beginning of calcification, crown completion, and apical closure of P_1-M_3. *Solid Bars* = Kronfeld; *open bars* = Garn *et al.* (From Stewart, 1963, Fig. 1, p. 267.)

Few people stop to realize that the average newborn baby is approximately 20 inches long and that during the first 18 years of life, the individual more than triples his length—if a male, to about 5½ to 6½ feet, and if a female, to about 5 to 5½ feet. The first 16 to 18 years are a period of very rapid growth (Fig. 4).

The bones of the human skeleton develop from a number of centers of ossification. It has been estimated that at about the 11th prenatal week there are approximately 806 centers of bone growth in the human skeleton. As the skeleton grows, these centers unite so that at birth there are some 450 centers and by adulthood these have united to form the 206 bones of the skeleton.

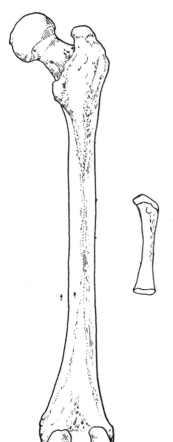

Fig. 4. Comparison of an adult femur with that of a new born infant (both rights).

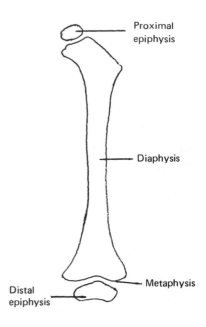

Proximal epiphysis

Diaphysis

Metaphysis

Distal epiphysis

Fig. 5. Femur of a subadult

A typical long bone will have three centers of ossification, one primary and two secondary.

Primary:

    Diaphysis or shaft—mid-portion of the bone

Secondary:

    2 Epiphyses or ends—the end portions of the bone

The layer of cartilage between an ossifying epiphysis and a diaphysis is known as an epiphyseal plate or disk. The metaphysis is the area of the diaphysis adjacent to the plate and in the region where growth in length takes place (Fig. 5).

All long bones (as well as metacarpals, metatarsals and phalanges) have an epiphysis at one and most have them at both ends. An epiphysis probably functions as a protective cap to the metaphysis or the active growing region of the bone (Grant: 1952). Long bones with epiphyses at both ends include:

Humerus
Radius
Ulna
Femur
Tibia
Fibula

The clavicle, first and second metacarpals and the first metatarsal occasionally have epiphyses at both ends.

Growth of Long Bones

Long bones do not grow at the same rate at both ends. In a long bone with two epiphyses, the first epiphysis to start ossification is at the end making the greatest contribution to growh in length. It is, however, the last to fuse with the diaphysis. Growth stops when the epiphyses unite with the diaphysis. It is not clearly understood exactly why the epiphyses unite when they do but it is involved with the endocrine system. Abnormalities can and do occur. Premature union of the epiphyses results in a dwarfed condition known as achondroplasia. Failure of union at the proper time results in giantism; however, this is seldom seen today, since union of the epiphyses can be stimulated by hormone therapy.

Ossification begins earlier in females, and the epiphyses unite earlier in females also, normally sóme two to three years sooner. This leads to a shorter period of growth in females and accounts for their smaller adult size when compared with males.

Epiphyses ossify from a single center with the cartilaginous epiphyseal disk being replaced by bone. Following complete ossification of the cartilaginous epiphyseal disk, union of the epiphysis and the diaphysis occurs.

All long bone shafts are ossified at birth as are many parts of the rest of the skeleton. Krogman (1962: 19-21) states that "at birth only six epiphyseal centers are present: head of humerus (proximal); condyle of femur (distal); condyle of tibia (proximal); talus, calcaneus and cuboid (tarsal bones in foot). The first three will unite with their respective shafts; the second three (as is true of all carpal bones of the hand and tarsal bones of the foot) will remain as discrete bones throughout life."

The early stages of the ossification of an epiphysis may be difficult to detect in skeletal remains. The epiphysis is then often no more than an amorphous lump. Because it is usually rounded and lacking in morphological detail, it may be discarded by the untrained excavator of burials as a pebble. As it gets larger it begins to take on its characteristic adult shape, and with more time morphological details can be identified. However, great care always must be taken in the recovery and identification of epiphyses.

Epiphyseal union—Each long bone of the body is made up of:

A diaphysis or shaft.

Two epiphyses or ends of bones (one at each end). Some bones have additional epiphyses (as greater and lesser trochanters of the femur).

A look at Chapter 2 of Krogman (1962) will show the reader that many lists of ages for epiphyseal unions are available. The following tables are from McKern and Stewart (1957), and represent some of the most recent data on epiphyseal union in males. In these tables, stages refer to:

0—open suture (no union)
1—one-quarter united or fused
2—one-half united or fused
3—three-quarters united or fused
4—completely united or fused

For additional coverage, refer to Greulich and Pyle (1959), Krogman (1962), and McKern and Stewart (1957).

TABLE 2.   THE AGE DISTRIBUTION FOR STAGES OF UNION FOR THE LONG BONE EPIPHYSES OF GROUP II (in %).   FROM McKERN & STEWART, 1957, TABLE 21, p. 45.

Upper Extremity

Age	No.	Humerus (prox.) Stages					Radius (dist.) Stages					Ulna (dist.) Stages				
		0	1	2	3	4	0	1	2	3	4	0	1	2	3	4
17-18	55	14	5	25	35	21	22	3	14	32	29	29	1	11	24	35
19	52	5	2	10	58	25	7	-	5	48	40	7	-	5	32	56
20	45	2	2	4	40	52	4	-	2	24	70	4	2	-	24	70
21	37			2	27	71				19	81				10	90
22	24				12	88				12	88				8	92
23	26				4	96					100					100
24+	136					100										
Total	375															

Lower Extremity

Age	No.	Femur (dist.) Stages					Tibia (prox.) Stages					Fibula (prox.) Stages				
		0	1	2	3	4	0	1	2	3	4	0	1	2	3	4
17-18	55	16	2	3	18	61	2	2	7	23	66	14	-	3	12	71
19	52	4	-	1	9	86	1	-	1	17	81	4	-	6	4	86
20	45			2	9	89				13	87			2	-	98
21	37				8	92				5	95				5	95
22	24					100				4	96					100
23	26										100					
24+	136															

TABLE 3. EPIPHYSIS ON ILIAC CREST: AGE DISTRIBUTION OF STAGES OF UNION (in %). FROM McKERN & STEWART, 1957, TABLE 22, p. 61.

Age	No.	Stages of Union				
		0	1	2	3	4
17	10	40	10	10	40	–
18	45	18	16	26	20	20
19	52	5	4	27	28	36
20	45	2	6	4	24	64
21	37	–	5	`8	13	74
22	24	–	–	4	4	92
23	26	–	–	–	–	100
Total	239					

Tooth Wear—In modern populations, tooth wear will not offer much help in aging and even in prehistoric populations it is of limited value. Much research needs to be conducted on dental attrition and its correlation with different foods and food preparation techniques.

Wear will not be marked to the same degree on all molars since they erupt at different ages. In other words the first molars are exposed to about twelve more years of mastication than the third molars and about six years more than the second molars. When an age determination is attempted that difference needs to be kept in mind. Again, it is also important to remember that all populations do not have the same rate of attrition and therefore the criteria of determining age from tooth wear in one population does not necessarily apply to another population. Unfortunately, all of the dentitions within a population do not wear at the same rate due to individual differences in diet and tooth structure. This severely limits the accuracy of age determination by this method and other criteria should be consulted whenever possible.

Following is D. R. Brothwell's age classification of wear on pre-medieval British teeth (Fig. 6).

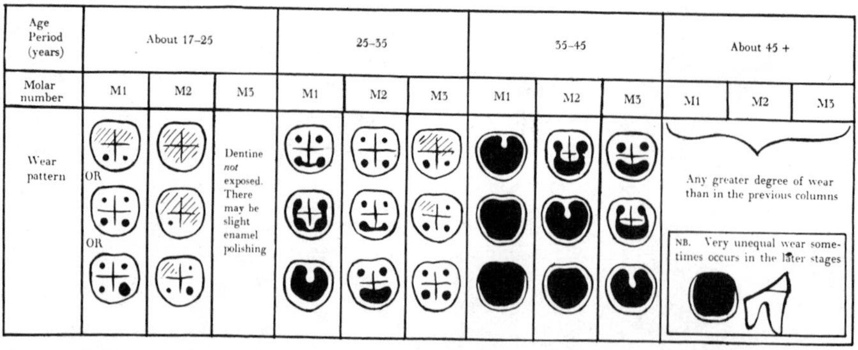

Fig. 6. A correlation of age at death with molar wear in pre-medieval British skulls (After Brothwell, 1965:69). Permission for reproduction granted by Dr. Donald R. Brothwell and the Trustees of the British Museum of Natural History.

18

Development of Osteoarthritis—Following the period of union of the epiphyses, that is, from the mid-20's onward, degenerative changes can be noted on and around the joint surfaces of many skeletons. Stewart (1958) has been interested in vertebral osteoarthritis as an aid in skeletal age identification. He studied the osteophytes (lipping) indicative of advancing age on the superior and inferior borders of each vertebral centrum and assigned a subjective rating scale running from 0 (no lipping) to 4+ (maximum lipping). His three figures (7, 8, 9) are summarized in Table 4.

TABLE 4. AGE ESTIMATION BY AMOUNT OF
VERTEBRAL OSTEOARTHRITIS.

Age	Amount of Lipping
20-30	Lipping develops rather slowly
30-40	Lipping intensifies
40-50	Lipping intensifies especially in lumbar region
50+	Lipping becomes quite pronounced

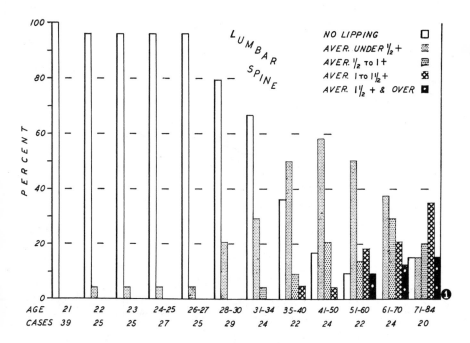

Fig. 7. Graph showing distribution of five categories of osteophytosis in 306 lumbar spines of white American males ranging in age from 21 to 84 years. (From Stewart, 1958, Fig. 1, p. 147.) Permission for reproduction granted by Dr. T. Dale Stewart.

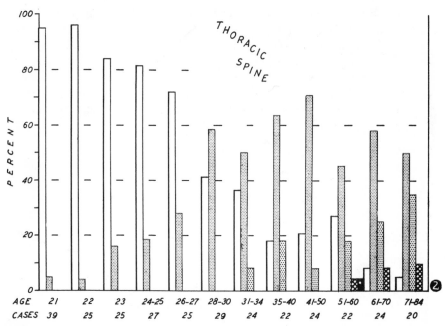

Fig. 8. Graph showing distribution of five categories of osteophytosis in 306 thoracic spines of white American males ranging in age from 21 to 84 years. (See Fig. 7 for categories.) (From Stewart, 1958, Fig. 2, p. 148.) Permission for reproduction granted by Dr. T. Dale Stewart.

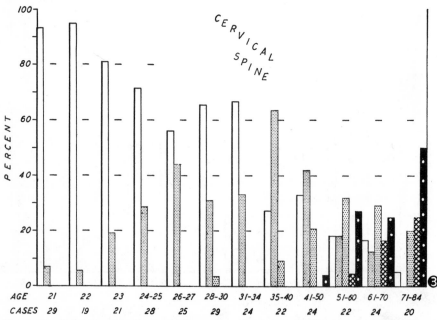

Fig. 9. Graph showing distribution of five categories of osteophytosis in 287 cervical spines of white American males ranging in age from 21 to 84 years. (See Fig. 7 for categories.) (From Stewart, 1958, Fig. 3, p. 149.) Permission for reproduction granted by Dr. T. Dale Stewart.

Sex Estimation

When a skeleton is discovered during excavation or observed in the laboratory, one of the first questions asked is "Is it male or female?" Many criteria for estimating the sex of a skeleton have been published in the anthropological and anatomical literature. As already stated, it is the intent here to present only a few basic criteria for sexing bones and to list references where additional information can be obtained.

The question still arises as to whether subadult skeletal material can be accurately sexed but the consensus is that any determination is little better than a guess. The secondary characteristics do not manifest themselves until puberty; thus it is impossible to judge the remains of children and adolescents because the means available relate to adult traits. Most of the techniques employed to sex subadult bones depend on X-ray taken in life. These techniques seldom apply to dried bones.

Selected references to sex determination of subadult skeletal material.

Boucher, B. J.
 1955 Sex difference in the foetal sciatic notch. Journal of Forensic Medicine, Vol. 2, pp. 51-54.
 1957 Sex differences in the foetal pelvis. American Journal of Physical Anthropology, Vol. 15, pp. 581-600.
Hunt, Edward E. and I. Gleiser
 1955 The estimation of age and sex of preadolescent children from bone and teeth. American Journal of Physical Anthropology, Vol. 13, pp. 479-487.
Imrie, J. A. and G. M. Wyburn
 1958 Assessment of age, sex and height from immature human bones. British Medical Journal, Vol. 1, pp. 128-131.
Reynolds, Earl L.
 1945 The bony pelvic girdle in early infancy. A roentgenometric study. American Journal of Physical Anthropology, Vol. 3, pp. 321-354.
 1947 The bony pelvis in prepuberal childhood. American Journal of Physical Anthropology, Vol. 5, pp. 165-200.

In general the sex differences in the adult long bones are a matter of size, typical male bones being longer and larger (more massive) than typical female bones (Krogman 1962: 143). In addition to simplify observation of size, measurement of the maximum diameter of the head of the humerus and of the femur is especially useful in sex determination.

Femur—Krogman (1962) gives a summary of data on measurements of the various long bones. Pearson (1917-19) gives the following information for maximum diameter in mm. of the head of the femur:

Female	Probably Female	Sex?	Probably Male	Male
X-41.5	41.5-43.5	43.5-44.5	44.5-45.5	45.5-X

Dwight (1905:22) gives the average maximum diameter of the femur head as 49.68 mm. for males and 43.84 mm. for females.

Humerus—Dwight's (1905:22) data on the diameter of the humeral head in mm.

	Vertical	Transverse
Male	48.76	44.66
Female	42.67	36.98

Stature Estimation

The estimation of living stature from the length of the long bones has long been of interest to the physical anthropologist. Some of the earliest scientific efforts along this line were undertaken in the middle and late 1800's on the basis of simple ratios.

It was not until the turn of the century (1899), with the development of mathematical regression equations, that Karl Pearson published a series of formulae for estimating living stature from the lengths of dried long bones. Lists of tables and formulae from the early work have been published by Krogman (1962: 153-187). Additional bibliographical references are given below. Because the most recent studies of Trotter and Gleser (1952 and 1958) are the most reliable, tables for whites and negroes are reproduced here (Table 3).

Various authors have demonstrated that estimation is complicated by racial differences between population samples. The racial affiliation of the sample must be known and the appropriate formulae or tables for that racial group used to estimate stature.

References to stature estimation formulae and tables.

Breitinger, E.

1938 Zur Berechnung der Körperhöhe aus den langen Gliedmassenknochen Anthropologischer Anzeiger, Vol. 14, pp. 249-274.

Dupertuis, C. W. and John A. Hadden, Jr.

1951 On the reconstruction of stature from long bones. American Journal of Physical Anthropology, Vol. 9, pp. 15-53.

Dwight, Thomas

1894 Methods of estimating the height from parts of the skeleton. Medical Records, New York, Vol. 65, pp.

Genovés, Santiago C.

1967 Proportionality of long bones and their relation to stature among Mesoamericans. American Journal of Physical Anthropology, Vol. 26, pp. 67-78.

Krogman, Wilton M.

1962 The human skeleton in Forensic Medicine. Charles C Thomas, Springfield, Illinois. (This has an entire chapter devoted to stature estimation.)

Manouvrier, L.

   1893 La determination de la taille d'apres les grands os des membres. Mém. Soc. d'Anthrop., Paris, 2 ser., Vol. 4, pp. 347-402.

Pearson, Karl

   1899 On the reconstruction of the stature of prehistoric races. Philos. Trans. Royal Society, London, Ser. A (Mathematical) Vol. 192, pp. 170-244.

Stevenson, Paul H.

   1929 On racial differences in stature long bone regression formulae, with special references to stature reconstruction formulae for the Chinese. Biometrika, Vol. 21, pp. 303-321.

Stewart, T. D. (Ed.)

   1952 Hrdlicka's practical anthropometry. Fourth edition. Wistar Institute of Anatomy and Biology, Philadelphia

Telkkä, Annti

   1950 On the prediction of human stature from the long bones. Acta Anat. Basle, Vol. 9, pp. 103-117.

Trotter, Mildred and Goldine C. Gleser

   1952 Estimation of stature from long bones of American Whites and Negroes. American Journal of Physical Anthropology, Vol. 10, pp. 463-514.

   1958 A re-evaluation of estimation of stature based on measurements of stature taken during life and of long bones after death. American Journal of Physical Anthropology, Vol. 16, pp. 79-123.

The following tables (5-9) are reproduced for the aid in calculating stature and in converting calculations in millimeters and centimeters to feet and inches.

TABLE 5. EXPECTED MAXIMUM STATURE* FROM LONG BONE LENGTHS (MAXIMUM) FOR AMERICAN WHITE MALES FROM TROTTER & GLESER, 1952, p. 496.

HUM	RAD	ULNA	STATURE		FEM	TIB	FIB	FEM + TIB
mm	mm	mm	cm	in **	mm	mm	mm	mm
265	193	211	152	59[7]	381	291	299	685
268	196	213	153	60[2]	385	295	303	693
271	198	216	154	60[5]	389	299	307	701
275	201	219	155	61	393	303	311	708
278	204	222	156	61[3]	398	307	314	716
281	206	224	157	61[6]	402	311	318	723
284	209	227	158	62[2]	406	315	322	731
288	212	230	159	62[5]	410	319	326	738
291	214	232	160	63	414	323	329	746
294	217	235	161	63[3]	419	327	333	753
297	220	238	162	63[6]	423	331	337	761
301	222	240	163	64[1]	427	335	340	769
304	225	243	164	64[5]	431	339	344	776
307	228	246	165	65	435	343	348	784
310	230	249	166	65[3]	440	347	352	791
314	233	251	167	65[6]	444	351	355	799
317	235	254	168	66[1]	448	355	359	806
320	238	257	169	66[4]	452	359	363	814
323	241	259	170	66[7]	456	363	367	821
327	243	262	171	67[3]	461	367	370	829
330	246	265	172	67[6]	465	371	374	837
333	249	267	173	68[1]	469	375	378	844
336	251	270	174	68[4]	473	379	381	852
339	254	273	175	68[7]	477	383	385	859
343	257	276	176	69[2]	482	386	389	867
346	259	278	177	69[5]	486	390	393	874
349	262	281	178	70[1]	490	394	396	882
352	265	284	179	70[4]	494	398	400	889
356	267	286	180	70[7]	498	402	404	897
359	270	289	181	71[2]	503	406	408	905
362	272	292	182	71[5]	507	410	411	912
365	275	294	183	72	511	414	415	920
369	278	297	184	72[4]	515	418	419	927
372	280	300	185	72[7]	519	422	422	935
375	283	303	186	73[2]	524	426	426	942
378	286	305	187	73[5]	528	430	430	950
382	288	308	188	74	532	434	434	957
385	291	311	189	74[3]	536	438	437	965
388	294	313	190	74[6]	540	442	441	973
391	296	316	191	75[2]	545	446	445	980
395	299	319	192	75[5]	549	450	449	988
398	302	321	193	76	553	454	452	995
401	304	324	194	76[3]	557	458	456	1003
404	307	327	195	76[6]	561	462	460	1010
408	309	330	196	77[1]	566	466	463	1018
411	312	332	197	77[4]	570	470	467	1026
414	315	335	198	78	574	474	471	1033

*The expected maximum stature should be reduced by the amount of .06 (age in years--30) cm to obtain expected stature of individuals over 30 years of age.

**The raised number indicates the numerator of a fraction of an inch expressed in eighths, thus 59[7] should be read 59 7/8 inches.

TABLE 6. EXPECTED MAXIMUM STATURE* FROM LONG BONE LENGTHS (MAXIMUM) FOR AMERICAN WHITE FEMALES FROM TROTTER & GLESER, 1952, p. 498.

HUM	RAD •	ULNA	STATURE		FEM	TIB	FIB	FEM + TIB
mm	mm	mm	cm	in **	mm	mm	mm	mm
244	179	193	140	55¹	348	271	274	624
247	182	195	141	55⁴	352	274	278	632
250	184	197	142	55⁷	356	277	281	639
253	186	200	143	56²	360	281	285	646
256	188	202	144	56⁶	364	284	288	653
259	190	204	145	57¹	368	288	291	660
262	192	207	146	57⁴	372	291	295	668
265	194	209	147	57⁷	376	295	298	675
268	196	211	148	58²	380	298	302	682
271	198	214	149	58⁵	384	302	305	689
274	201	216	150	59	388	305	309	696
277	203	218	151	59⁴	392	309	312	704
280	205	221	152	59⁷	396	312	315	711
283	207	223	153	60²	400	315	319	718
286	209	225	154	60⁵	404	319	322	725
289	211	228	155	61	409	322	326	732
292	213	230	156	61³	413	326	329	740
295	215	232	157	61⁶	417	329	332	747
298	217	235	158	62²	421	333	336	754
301	220	237	159	62⁵	425	336	340	761
304	222	239	160	63	429	340	343	768
307	224	242	161	63³	433	343	346	776
310	226	244	162	63⁶	437	346	349	783
313	228	246	163	64¹	441	350	353	790
316	230	249	164	64⁵	445	353	356	797
319	232	251	165	65	449	357	360	804
322	234	253	166	65³	453	360	363	812
324	236	256	167	65⁶	457	364	366	819
327	239	258	168	66¹	461	367	370	826
330	241	261	169	66⁴	465	371	373	833
333	243	263	170	66⁷	469	374	377	840
336	245	265	171	67²	473	377	380	847
339	247	268	172	67⁶	477	381	384	855
342	249	270	173	68¹	481	384	387	862
345	251	272	174	68⁴	485	388	390	869
348	253	275	175	68⁷	489	391	394	876
351	255	277	176	69²	494	395	397	883
354	258	279	177	69⁵	498	398	401	891
357	260	282	178	70¹	502	402	404	898
360	262	284	179	70⁴	506	405	407	905
363	264	286	180	70⁷	510	409	411	912
366	266	289	181	71³	514	412	414	919
369	268	291	182	71⁵	518	415	418	927
372	270	293	183	72	522	419	421	934
375	272	296	184	72⁴	526	422	425	941

*The expected maximum stature should be reduced by the amount of .06 (age in years--30) cm to obtain expected stature of individuals over 30 years of age.

**The raised number indicates the numerator of a fraction of an inch expressed in eighths, thus 55¹ should be read 55 1/8 inches.

TABLE 7. EXPECTED MAXIMUM STATURE* FROM LONG BONE LENGTHS (MAXIMUM) FOR AMERICAN NEGRO MALES FROM TROTTER & GLESER, 1952, p. 497.

HUM	RAD	ULNA	STATURE		FEM	TIB	FIB	FEM + TIB
mm	*mm*	*mm*	*cm*	*in** *	*mm*	*mm*	*mm*	*mm*
276	206	223	152	59⁷	387	301	303	704
279	209	226	153	60²	391	306	308	713
282	212	229	154	60⁵	396	310	312	721
285	215	232	155	61	401	315	317	730
288	218	235	156	61³	406	320	321	739
291	221	238	157	61⁶	410	324	326	747
294	224	242	158	62²	415	329	330	756
297	226	245	159	62⁵	420	333	335	765
300	229	248	160	63	425	338	339	774
303	232	251	161	63³	430	342	344	782
306	235	254	162	63⁶	434	347	349	791
310	238	257	163	64¹	439	352	353	800
313	241	260	164	64⁵	444	356	358	808
316	244	263	165	65	449	361	362	817
319	247	266	166	65³	453	365	367	826
322	250	269	167	65⁶	458	370	371	834
325	253	272	168	66¹	463	374	376	843
328	256	275	169	66⁴	468	379	381	852
331	259	278	170	66⁷	472	383	385	861
334	262	281	171	67³	477	388	390	869
337	264	284	172	67⁶	482	393	394	878
340	267	287	173	68¹	487	397	399	887
343	270	291	174	68⁴	491	402	403	895
346	273	294	175	68⁷	496	406	408	904
349	276	297	176	69²	501	411	413	913
352	279	300	177	69⁵	506	415	417	921
356	282	303	178	70	510	420	422	930
359	285	306	179	70¹	515	425	426	939
362	288	309	180	70⁷	520	429	431	947
365	291	312	181	71²	525	434	435	956
368	294	315	182	71⁵	529	438	440	965
371	297	318	183	72	534	443	445	974
374	300	321	184	72⁴	539	447	449	982
377	302	324	185	72⁷	544	452	454	991
380	305	327	186	73²	548	456	458	1000
383	308	330	187	73⁵	553	461	463	1008
386	311	333	188	74	558	466	467	1017
389	314	336	189	74³	563	470	472	1026
392	317	340	190	74⁶	567	475	476	1034
395	320	343	191	75²	572	479	481	1043
398	323	346	192	75⁵	577	484	486	1052
401	326	349	193	76	582	488	490	1061
405	329	352	194	76³	586	493	495	1069
408	332	355	195	76⁶	591	498	499	1078
411	335	358	196	77¹	596	502	504	1087
414	337	361	197	77⁴	601	507	508	1095
417	340	364	198	78	605	511	513	1104

*The expected maximum stature should be reduced by the amount of .06 (age in years--30) cm to obtain expected stature of individuals over 30 years of age.

**The raised number indicates the numerator of a fraction of an inch expressed in eighths, thus 59⁷ should be read 59 7/8 inches.

TABLE 8. EXPECTED MAXIMUM STATURE* FROM LONG BONE LENGTHS (MAXIMUM) FOR AMERICAN NEGRO FEMALES FROM TROTTER & GLESER, 1952, p. 499.

HUM	RAD	ULNA	STATURE		FEM	TIB	FIB	FEM + TIB
mm	mm	mm	cm	in **	mm	mm	mm	mm
245	165	195	140	55[1]	352	275	278	637
248	169	198	141	55[4]	356	279	28[]	645
251	173	201	142	55[7]	361	283	286	653
254	176	204	143	56[2]	365	287	290	661
258	180	207	144	56[6]	369	291	294	669
261	184	210	145	57[1]	374	295	298	677
264	187	213	146	57[4]	378	299	302	685
267	191	216	147	57[7]	383	303	306	693
271	195	219	148	58[2]	387	308	310	701
274	198	222	149	58[5]	391	312	314	709
277	202	225	150	59	396	316	318	717
280	205	228	151	59[4]	400	320	322	724
284	209	231	152	59[7]	405	324	326	732
287	213	235	153	60[2]	409	328	330	740
290	216	238	154	60[5]	413	332	334	748
293	220	241	155	61	418	336	338	756
297	224	244	156	61[3]	422	340	342	764
300	227	247	157	61[6]	426	344	346	772
303	231	250	158	62[2]	431	348	350	780
306	235	253	159	62[5]	435	352	354	788
310	238	256	160	63	440	357	358	796
313	242	259	161	63[3]	444	361	362	804
316	245	262	162	63[6]	448	365	366	812
319	249	265	163	64[1]	453	369	370	820
322	253	268	164	64[5]	457	373	374	828
326	256	271	165	65	462	377	378	836
329	260	274	166	65[3]	466	381	382	843
332	264	277	167	65[6]	470	385	386	851
335	267	280	168	66[1]	475	389	390	859
339	271	283	169	66[4]	479	393	394	867
342	275	286	170	66[7]	484	397	398	875
345	278	289	171	67[3]	488	401	402	883
348	282	292	172	67[6]	492	406	406	891
352	285	295	173	68[1]	497	410	410	899
355	289	298	174	68[4]	501	414	414	907
358	293	301	175	68[7]	505	418	418	915
361	296	304	176	69[2]	510	422	422	923
365	300	307	177	69[5]	514	426	426	931
368	304	310	178	70[1]	519	430	430	939
371	307	313	179	70[4]	523	434	434	947
374	311	316	180	70[7]	527	438	438	955
378	315	319	181	71[2]	532	442	442	963
381	318	322	182	71[5]	536	446	446	970
384	322	325	183	72	541	450	450	978
387	325	328	184	72[4]	545	454	454	986

*The expected maximum stature should be reduced by the amount of .06 (age in years--30) cm to obtain expected stature of individuals over 30 years of age.

**The raised number indicates the numerator of a fraction of an inch expressed in eighths, thus 55[1] should be read 55 1/8 inches.

TABLE 9

Calculation of stature (in cm) from long bones[1] for Mesoamericans
(From Genovés 1966, Table 19, p. 41)*

Males:

All bones: Stature = 2.52 Rad - 0.07 Ulna + 0.44 Hum + 2.98 Fib - 0.49 Tib + 0.68 Fem + 951.13 $\pm$ 26.14

Femur: Stature = 2.26 Fem + 663.79 $\pm$ 34.17

Tibia: Stature = 1.96 Tib + 937.52 $\pm$ 28.15

Females:

All bones:[2] Stature = 8.66 Rad - 7.37 Ulna + 1.25 Tib - 0.93 Fem + 966.74 $\pm$ 28.12

Femur: Stature = 2.59 Fem + 497.42 $\pm$ 38.16

Tibia: Stature = 2.72 Tib + 637.81 $\pm$ 35.13

---

1 Subtract 2.5 cm to obtain the stature while alive.

2 Working at the level of precision of the Gamma 30 confirmed that, in this case, the humerus contributes practically nothing, and therefore was automatically omitted.

* Genoves says (1967) Table 14 was incorrectly printed in the journal with errors of several figures. Table 9 above has been corrected. (Personnel Communication 1971).

II. THE SKULL OR CRANIUM

Skull is the term used to denote the bony supporting framework of the head. It presents the most complex unit of the skeleton because of its function to protect the brain, one of the most vital parts of the body, as well as the organs of sight, hearing, smell, mastication and taste. Technically the cranium is the skull minus lower jaws; the calvarium is the cranium minus face; and the calva is the calvarium minus base.

How to handle the skull:

One of the first things a student must learn is how to handle a skull. Taken in general the skull is seen as spheroidal in shape with a smooth top, somewhat compressed from side to side and having an uneven base. To some it resembles a bowling ball and there is a compulsion to insert one's fingers into the eye orbits or sockets. *This should never be done.* The bones of the medial wall of the eye orbit are paper thin and are easily broken with slight pressure.

Always

1. Hold the skull with both hands and make sure you have a firm grip on on the skull at all times.
2. Place on a bean bag or donut ring when placed on a flat surface. It is round and may roll off a table (see introduction).
3. Care should be taken against breaking teeth when putting jaws together.

Never

1. Never insert the fingers into the eye orbits.
2. Never hold a skull by the zygomatic arch. Often in archaeological specimens this area is fragile or cracked and will break easily.
3. One should not try to hold the skull by inserting the fingers into the foramen magnum, the large hole on the base of the skull through which the spinal column passes.
4. Never place on a flat surface (table) unless on a bean bag, donut ring, sand box or some other secure foundation.

Bones of the Skull

There are 22 bones in the skull (Figs. 10-12) (six unpaired and eight paired) plus six ear bones (three in each ear), a total of 28 (Table 10). Although the hyoid is often grouped with the skull it is usually not considered a part of the skull.

BONES OF

PARTS OF

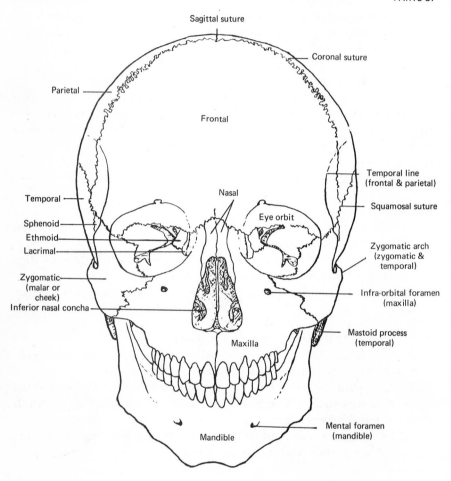

Fig. 10. Frontal (Anterior) view

BONES OF

PARTS OF

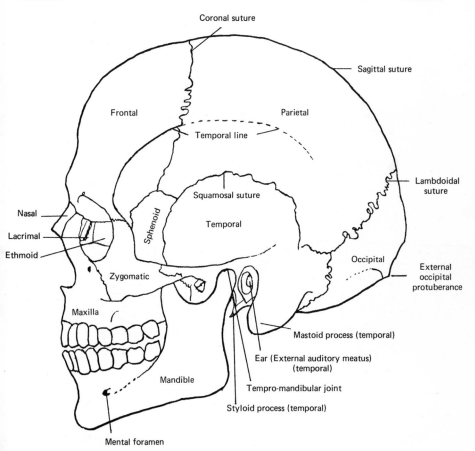

Coronal suture

Sagittal suture

Frontal

Parietal

Temporal line

Lambdoidal suture

Squamosal suture

Nasal

Sphenoid

Temporal

Lacrimal

Ethmoid

Zygomatic

Occipital

External occipital protuberance

Maxilla

Mastoid process (temporal)

Ear (External auditory meatus) (temporal)

Mandible

Tempro-mandibular joint

Styloid process (temporal)

Mental foramen

Fig. 11. Lateral view

BONES OF

PARTS OF

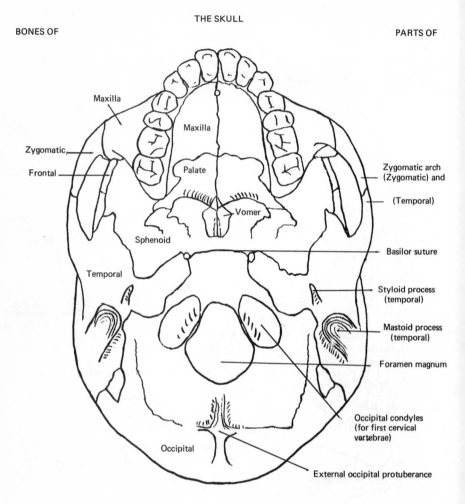

Maxilla

Maxilla

Zygomatic

Frontal

Palate

Vomer

Sphenoid

Temporal

Zygomatic arch
(Zygomatic) and

(Temporal)

Basilor suture

Styloid process
(temporal)

Mastoid process
(temporal)

Foramen magnum

Occipital condyles
(for first cervical
vertebrae)

Occipital

External occipital protuberance

Fig. 12. Base view

TABLE 10. BONES OF THE SKULL

Name of Bone	Single	Paired
CRANIUM		
Frontal	1	
Parietal		2
Occipital	1	
Temporal		2
Sphenoid	1	
Maxilla		2
Nasal		2
Zygomatic (malar or cheek)		2
Mandible	1	
Ethmoid	1	
Lacrimal		2
Palate		2
Vomer	1	
Inferior nasal concha		2
EAR		
Malleus		2
Incus		2
Stapes		2
ASSOCIATED		
Hyoid	1	

Sutures:

The bones of the skull come together or unite along special serrated and interlocking joints known as *SUTURES.* The sutures are irregular linear gaps before the age of 17 but often unite in old age and eventually become obliterated. It should be noted that ossification of the sutures occurs inside the skull (endocranially) first and proceeds toward the outside (ectocranially).

Most sutures take their names from the two bones of the skull that go together to form the suture. However, there are five notable exceptions to this rule:

Suture name	Location
Coronal	Between frontal and parietals
Sagittal	Between the two parietals
Lambdoidal	Between parietals and occipital
Basilar	Between occipital and sphenoid
Squamosal	Between temporal and parietal

BONES OF THE CRANIAL VAULT

Frontal Bone: SINGLE (May be divided on midline by Metopic suture) (Fig. 13).

Lying under the forehead and forming the upper edges of the orbits or eye sockets it articulates with the parietals, sphenoid, ethmoid, lacrimal, zygomatic, maxilla, and nasal bones.

EXTERNAL SURFACE   CEREBRAL SURFACE

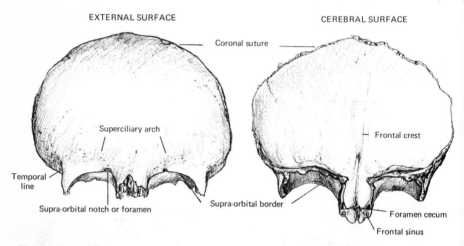

Fig. 13. Frontal bone

ANATOMICAL CHARACTERISTICS OF IMPORTANCE IN IDENTIFICATION:

Exterior surface

Superciliary arches (brow ridges)—curved projections above orbits.

Supra-orbital notch or foramen—a hole or notch, or both, in the supra-orbital border for the passage of the supra-orbital nerve or artery.

Supra-orbital border—upper edge of eye orbit; often used in sex determination.

Temporal line—edge of attachment of the temporal fascia and the lower end of the temporal muscle.

Cerebral surface

Frontal crest—a crest on anterior part of bone that terminates in the foramen cecum.

Foramen cecum—between frontal crest and crista galli of the ethmoid (sometimes transmits a vein from nasal cavity to superior sagittal sinus).

BONES OF SIMILAR SHAPE WHERE CONFUSION MAY ARISE:

The following cranial bones are all flat: *parietals, occipital* and *temporals.* In the postcranial skeleton, the *scapula* and *innominates* also have a broad flat surface.

Only the frontal bone forms the upper edge of the orbits.

Smooth areas of the frontal sinuses are located on the cerebral surface between the orbits.

If you find a sinus area on a bone remember that large sinuses occur on only the frontal, maxilla, and sphenoid.

The cerebral surface of the roof of the orbits has a very characteristic rough surface with furrows and ridges corresponding to the convolutions of the brain. This combination of furrows and ridges is found only on the frontal bone.

The temporal lines are well marked on the external surface of the frontal.

The frontal crest terminates in the foramen cecum located on the anterior border of the frontal bone between the orbits.

The posterior border of the frontal bone is the coronal suture.

Parietal Bone: PAIRED (Fig. 14)

Forming a large part of the roof and sides of the cranium, the parietal bone is interposed between the frontal and occipital bones. It articulates with the occipital, frontal, sphenoid, temporal and the other parietal.

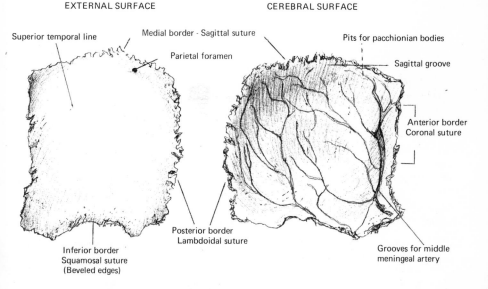

Fig. 14. Left Parietal

ANATOMICAL CHARACTERISTICS OF IMPORTANCE FOR IDENTIFICATION:

Exterior surface
Superior temporal line.

Parietal foramen—found in the posterior one-third near the sagittal suture. They usually occur in pairs but may not be present and are seldom multiple.

Frontal (anterior) border—the border along the coronal suture that is slightly concave and is less serrated than either the sagittal or lambdoidal sutures.

Medial border—identified by the sagittal suture, which is more serrated than the coronal suture and is relatively straight.

Posterior border—the lambdoidal suture is convex and deeply serrated.

Inferior border or squamosal border—is one of the few with a beveled edge (Fig. 15). It presents a series of different patterns, and in general is concave (to fit the convex temporal bone). Divisions of the squamosal border are as follows:

Area	Pattern	Articulates with
Anterior	Thin and beveled	Sphenoid
Medial	Thin and beveled	Temporal
Posterior	Thick and serrated	Mastoid portion of temporal

Fig. 15.

Cerebral surface

Grooves for the middle meningeal vessels always run up and back from the sphenoid angle. Note that there is a heavy groove just posterior to the coronal (anterior) border.

The sagittal groove (for the superior sagittal sinus) runs along the inside of the sagittal suture.

Pits for the pacchionian bodies are sometimes seen along the sides of the sagittal groove, especially in its anterior one-half.

BONES OF SIMILAR SHAPE WHERE CONFUSION MAY ARISE:

The following cranial bones are all flat: *frontal, occipital* and *temporal.* In the postcranial skeleton, the *scapula* and *innominate* also have a broad flat surface.

INFORMATION OF IMPORTANCE IN IDENTIFICATION:

Always check the edges of the bone carefully. If the edges are serrated it will be a cranial bone.

The parietals are shallow bowl-shaped bones with sutures on all four edges.

Check for the lines of the middle meningeal vessels (cerebral surface). These occur on only parietals and temporals and sometimes on the posterior edge of the frontal.

Side Identification:

Always try to place any bone as it is in your body. This will help in orientating the bone.

The concave beveled suture is always down or away from the midline.

When the parietal foramina are present, they always occur in the posterior third of the sagittal suture near the midline.

The most deeply serrated suture is the lambdoidal, which is always posterior.

The grooves for the middle meningeal vessels always run up and back. If you are holding the parietal in a position so that it is in close relationship to your own parietal and the grooves for the middle meningeal vessels are running up and forward, you must change the bone to the opposite side of the body to have it in proper orientation.

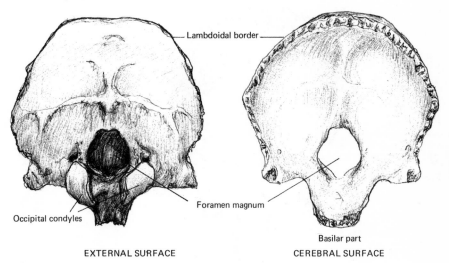

EXTERNAL SURFACE

CEREBRAL SURFACE

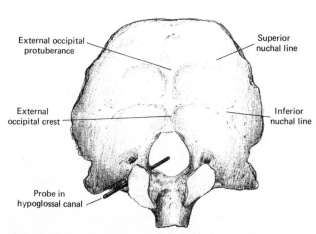

Fig. 16 Occipital bone

Occipital Bone: SINGLE (Fig. 16).

Situated in the posterior and inferior part of the cranium, it is connected by sutures with the two parietals (lambdoidal suture), the two temporals (along the mastoid margin) and the sphenoid (basilar "suture"). This is the only bone of the skull which articulates with the postcranial skeleton. The articulation is through the occipital condyles with the first cervical vertebra (atlas). In a few individuals the occipital may articulate with the dens of the epistropheus (axis). The largest hole in the skull, the foramen magnum, through which the spinal cord enters the skull, is located on the inferior surface between the occipital condyles. The occipital develops in four parts (Fig. 17).

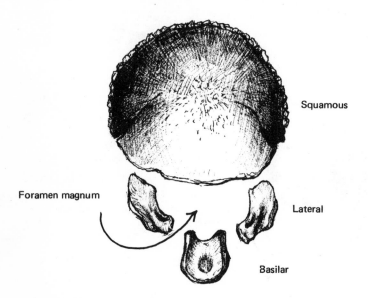

Squamous

Foramen magnum

Lateral

Basilar

Fig. 17. Occipital bone at birth

Four parts

Basilar—anterior to foramen magnum; articulates at the basilar suture with the sphenoid and is the smallest of the four parts.

Two laterals—form the sides of the foramen magnum; they articulate with the temporal bones. The major part of the occipital condyles is located on the lateral portions.

Squamous—posterior to the foramen magnum; the largest of the four parts, composed of the interparietal and supra-occipital portions.

ANATOMICAL CHARACTERISTICS OF IMPORTANCE IN IDENTIFICATION:

External surface

External occipital protuberance—midway between the superior border (lambdoidal suture) and the posterior margin of the foramen magnum.

Superior nuchal line—arching laterally on each side from the external occipital protuberance toward the lateral angles of the bone.

Inferior nuchal line—runs laterally from the middle of the external occipital crest to the jugular process.

External occipital crest—medium ridge running from external occipital protuberance to foramen magnum.

Occipital condyles—articular surfaces for the first cervical vertebra.

Cerebral surface

Internal occipital crest—note that the internal surface is concave and marked by two grooved ridges that cross each other at the internal occipital protuberance and divide the surface into four parts. The internal occipital crest is a median ridge which divides as it approaches the foramen magnum and becomes less defined.

BONES OF SIMILAR SHAPE WHERE CONFUSION MAY ARISE:

The following cranial bones are all flat: *frontal, parietals* and *temporals.* In the postcranial skeleton the *scapula* and *innominates* also have a broad flat surface.

INFORMATION OF IMPORTANCE IN IDENTIFICATION:

If you have a flat cranial bone, look first for the foramen magnum.

Articular surfaces occur on only three cranial bones: (1) occipital with the occipital condyles, (2) temporal with the temporomandibular joint, and (3) the condyles of the mandible.

The occipital condyles are convex, the temporomandibular articular surfaces are concave.

The lambdoidal suture is deeply serrated.

The internal configuration exhibits two crests that divide the surface into four parts, which is a feature distinguishing it from the frontal bone.

Temporal Bone: PAIRED (Fig. 18)

Located at the side and base of the cranium, the temporal bone is below the parietal, posterior to the sphenoid and anterior to the occipital. The organs of hearing and the articulation for the mandible are contained in this bone. Each one also articulates with a zygomatic (malar or cheek) bone through the zygomatic arch.

This bone has three parts:

Squamous—flat anterior and superior part which articulates with the sphenoid and through a beveled suture with the parietal bones; includes the zygomatic process.

Mastoid—that area behind the ear opening (external auditory meatus) that is thick and conical and projects downward.

Petrous—medial to the last two, lies in the base of the skull and contains the inner ear.

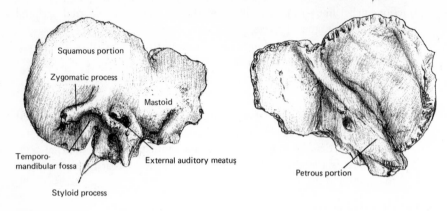

EXTERNAL SURFACE

Squamous portion

Zygomatic process

Mastoid

Temporo-
mandibular fossa

External auditory meatus

Styloid process

CEREBRAL SURFACE

Petrous portion

Fig. 18. Left Temporal bone

ANATOMICAL CHARACTERISTICS OF IMPORTANCE IN IDENTIFICATION:

Exterior surface

Zygomatic process—that portion of the bone extending forward and articulating with the malar bone to form the zygomatic arch.

Mandibular fossa (Temporomandibular joint)—the articular surface for the condyle of the mandible. It is immediately anterior to the ear hole (external auditory meatus) and just below the posterior end of the zygomatic arch.

External auditory meatus—the ear opening. When the bone is in proper anatomical position it is posterior to the zygomatic process but anterior to the mastoid process.

Mastoid process—a cone-shaped projection pointing downward behind the external auditory meatus. It is the bony structure that can be felt back of the ear.

Styloid process—is on the base of the skull and medial to the mandibular fossa, external auditory meatus, and mastoid process. It is a slender cylindrical spur that projects down and forward and gives attachments to muscles, especially those involved in speech.

Cerebral surface

Petrous portion—projects almost at a right angle from the squamous and mastoid portions; situated to the medial side of these parts and participates in the formation of the base of the skull. The petrous portion fills the area in the base of the skull between the basilar and lateral portions of the occipital bone and the greater wings of the sphenoid. It is functionally the most important part of the temporal bone, for it surrounds the essential part of the organs of hearing.

Grooves for the middle meningeal vessels—can often be seen along the anterior and superior borders.

This is a complicated bone and when fragmentary can to the unexperienced resemble almost any bone. Bones often confused with the temporal are the *parietal* (because of the beveled suture), the *occipital* and the *frontal*. Postcranial bones having flat surfaces are the *scapula* and *innominate*.

Side identification:

When held in anatomical order the zygomatic process always points forward and the mastoid process downward.

The mandibular fossa is always in front of the ear opening (external auditory meatus).

The squamosal border is beveled with a sharp edge and is always superior.

The styloid process always points 'downward.

The mastoid process is always posterior to the ear opening (external auditory meatus).

Sphenoid Bone: SINGLE (Fig. 19)

A very irregular bone which helps form the floor and sides of the cranial vault. Composed of a body, two pairs of lateral expansions called greater and lesser wings, and a pair of processes that project downward (the pterygoid processes), it is in general a U-shaped bone. It articulates with the occipital, parietals, frontal, ethmoid, temporals, palatine, vomer, zygomatics, and sometimes with the maxillae.

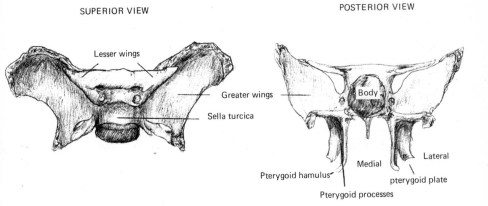

Fig. 19. Sphenoid

External surface

Greater wings—plates of bone that extend laterally from the body, and bend superiorly near their tips. They form the portion of the skull between the temporals and frontal.

Pterygoid processes—project downward from the junction of the body and greater wings. These are best seen as the structures at the posterior end of the tooth row. Each projection consists of two plates (the pterygoid plates). The lateral one is shorter and broader and the medial is longer and narrower. They unite in front but diverge behind.

Cerebral surface

Lesser wings—two thin triangular plates of bone extending laterally and almost horizontally from the anterior portion of the body. They are on the cerebral surface immediately anterior to the greater wings.

Sella turcica—a depression on the superior surface of the body. The pituitary gland is housed in this depression posterior to the lesser wings.

BONES OF SIMILAR SHAPE WHERE CONFUSION MAY ARISE:

Because of its complicated structure, it is possible to confuse this bone with most of the cranial bones. Those most frequently confused are the *temporal, parietal, occipital* and *frontal.*

INFORMATION OF IMPORTANCE IN IDENTIFICATION:

This bone has a number of sharp projections.
It has many foramina.
Most of the bone is fragile.
The body contains the sphenoidal sinus.
Remember that sinuses occur in only the frontal, maxilla, zygomatic, and sphenoid.

BONES OF THE FACE

Maxilla (Upper Jaw): PAIRED (Fig. 20).

One of the largest bones of the face, it supports the upper teeth and helps form the orbits, the hard palate and the nasal fossa. It is divided into a body and four processes which are listed below. The two maxillae articulate with each other and with the frontal, nasals, lacrimals, ethmoid, palate bones, vomer, zygomatics and inferior nasal conchae. In some cases the maxillae may articulate with the sphenoid.

ANATOMICAL CHARACTERISTICS OF IMPORTANCE IN IDENTIFICATION:

Body—this is the main portion of the bone. It is hollow because of the maxillary sinus.

Infra-orbital foramen—located just below the inferior margin of the eye orbits. Terminal branches of the infra-orbital nerves and blood vessels emerge here.

Anterior nasal spine—this is a sharp anterior projection along the midline at the base of the nose.

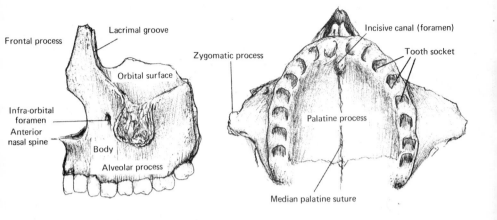

Fig. 20. Maxilla

LEFT MAXILLA

RIGHT AND LEFT MAXILLAE
INFERIOR VIEW
(TEETH REMOVED)

Frontal process—a slender portion of the bone that rises from the body and helps to form the lateral wall of the nasal fossa. It articulates with the frontal bone and with the lacrimal bone forms the lacrimal groove on the anterior medial surface of the eye socket.

Zygomatic process—below the eye socket; it is a lateral projection from the body that articulates along a diagonal suture with the malar or zygomatic bone.

Palatine process—forms the roof of the mouth. It projects medially from the body and joins the process from the opposite side at the midline. This process forms three-fourths of the hard palate.

Incisive foramen—in the midline behind the central incisors is a large foramen or canal formed by the right and left maxillae.

Alveolar process—that portion of the bone that extends inferiorly from the body and holds the upper teeth. There are usually eight tooth sockets in each maxilla arranged in a U-shaped arch.

BONES OF SIMILAR SHAPE WHERE CONFUSION MAY ARISE:

The *mandible* because of the teeth and/or tooth sockets.

Side identification:

This is not a difficult bone to side if oriented in anatomical position.

Remember that the major sinuses are found in the maxilla. Other bones with sinuses are the frontal, sphenoid and zygomatic.

The incisive foramen is along the midline.

The zygomatic process extends away from the midline.

Nasal Bone: PAIRED (Fig. 21)

This is a small bone that forms the bridge of the nose. Each bone is thicker and narrower superiorly and thinner and wider inferiorly, and presents two surfaces and four borders. The nasal bones articulate with each other and with the frontal, the two maxillae, and the ethmoid.

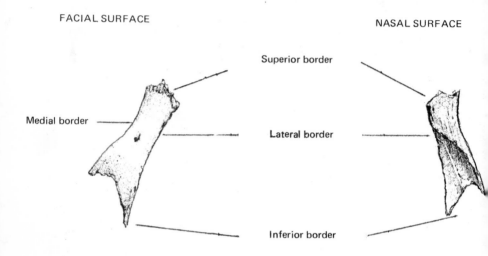

Fig. 21. Left Nasal bone

ANATOMICAL CHARACTERISTICS OF IMPORTANCE IN IDENTIFICATION:

Medial border—a finely serrated articular border where the two nasal bones come together. Of the three articular edges this is intermediate in length. Forms the internasal suture.

Lateral border—the longest of the three articular edges and is the border that joins the maxilla. It is more deeply serrated than the medial border but not as serrated as the superior border. Forms the naso-maxillary suture.

Superior border—the shortest and most deeply serrated of the three articular borders. Forms the nasofrontal suture.

Inferior border—presents the only nonarticular border. The bone is thin here and presents a sharp edge.

BONES OF SIMILAR SHAPE WHERE CONFUSION MAY ARISE:

Possibly the *vomer*.

Side identification:

Remember that there are three articular borders and one nonarticular.

The longest articular border is the lateral edge (nasiomaxillary suture).

When held in approximate anatomical position with the nonarticular surface down or away from you, the nasiomaxillary suture will be on the same side the bones come from.

The finely serrated articular border is the medial border.

Zygomatic (Malar or Cheek) Bone: PAIRED (Fig. 22)

This bone forms the prominence of the cheek and can be felt under the skin just below and lateral to the eye socket. It articulates with the zygomatic portion of the temporal bone to form the zygomatic arch. It also articulates with the maxilla, frontal and sphenoid. The zygomatic bone can be called by either of two other terms, malar or cheek bone, but the term zygomatic should be learned as it is the most used and will aid in remembering the zygomatic arch.

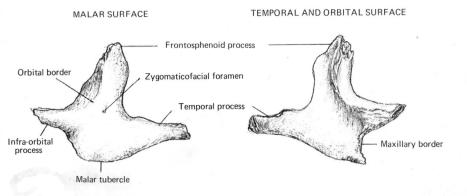

MALAR SURFACE   TEMPORAL AND ORBITAL SURFACE

Frontosphenoid process

Orbital border

Zygomaticofacial foramen

Temporal process

Infra-orbital process

Maxillary border

Malar tubercle

Fig. 22. Left Zygomatic bone

ANATOMICAL CHARACTERISTICS OF IMPORTANCE IN IDENTIFICATION:

Infra-orbital process—the anterior portion just below the eye socket which articulates with the maxilla.

Frontosphenoid process—the superior projection of the bone which forms the lateral edge of the eye socket. On the surface of the bone just below this process, the zygomaticofacial foramen is sometimes found.

Zygomaticofacial foramen—the name applied to one and sometimes two or more small holes on the anterior surface of the bone. They transmit the zygomaticofacial nerves and vessels.

Temporal process—the posterior projection that articulates with the temporal bone to form the zygomatic arch.

Malar tubercle—the blunt, rounded, inferior angle of the bone.

The *frontal* and *maxillae* because of the edges of the eye orbits. In the post-cranial skeleton some students confuse the edges of the *scapula* with the zygomatic.

Side identification:

The infra-orbital process ends in a sharp point and is always below the eye orbit and points toward the nose.

The maxillary border is deeply serrated and angles from the edge of the eye orbit (infra-orbital process) laterally.

This bone is thin and has an anterior and a temporal surface.

Mandible: SINGLE (Fig. 23)

The mandible or lower jaw is the largest and strongest bone of the face. It is the most movable bone of the skull, articulating through the condyles with the temporal bones at the temporomandibular joints. The bone supports the lower teeth which meet those of the maxilla at the occlusal plane.

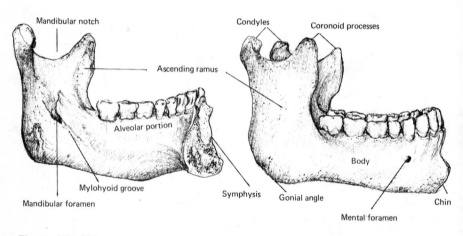

Fig. 23. Mandible

ANATOMICAL CHARACTERISTICS OF IMPORTANCE IN IDENTIFICATION:

Body—the anterior and horizontal portion of the bone that is shaped like a horseshoe. It is sometimes known as the horizontal ramus.

Ascending ramus—consisting of a right and a left, these are the broad and vertical projections that terminate posteriorly with the condyles for articulation with the temporal bones and anteriorly with the nonarticular coronoid processes.

Condyle—oval in form, and through the articular surface (the head) it joins the temporal bone. It is the posterior projection of the ascending ramus.

Coronoid process—the anterior superior projection of the ascending ramus. A flattened and triangular portion of the bone gives attachments to the temporal and masseter muscles. This projection is separated from the condyle by the mandibular notch.

Mandibular notch—a deep notch between each anterior coronoid process and the posterior condyles (condyloid process).

Symphysis—on the midline of the body may be seen a faint line where the two original separate halves of the bone united. The two halves usually unite just prior to birth.

Mental foramen—on the external surface of the body and just below the premolar teeth. It is approximately midway between the upper and lower border and transmits the mental nerve and vessels.

Mandibular foramen—two large holes on the medial surface and approximately in the middle of the ascending ramus. The mandibular nerve enters the bone here to innervate the teeth in the lower jaw.

Mylohyoid groove—runs obliquely downward and forward from the mandibular foramen and holds the mylohyoid nerve and artery. Sometimes this groove is bridged over with bone to form a mylohyoid bridge.

Gonial angle—the angle formed by the meeting of the thick posterior border of the horizontal ramus and the inferior border of the ascending ramus. It is usually everted but can occur as a straight or inverted angle.

Chin—a structure found only in man, and is the anterior projection of the inferior border of the body or horizontal ramus. (This occurs in fossil man only in *Homo sapiens* and dates back about 35,000-40,000 years with the appearance of Cro-Magnon man).

Alveolar process—the superior border of the body or horzontal ramus which is hollowed out into sockets for the teeth. In old age when the teeth are lost there may be resorption.

BONES OR SIMILAR SHAPE WHERE CONFUSION MAY ARISE:

Maxilla because of the teeth. Remember that there are no sinuses at the base of the mandibular teeth as there are in the maxilla. The base of the alveolar portion is expanded in the maxilla while it is narrow dense bone in the mandible.

Side identification:

Although this is a single bone it is frequently broken and the side must be identified.

The teeth are always superior and the ascending ramus posterior.

The mental foramen is on the external surface and the mandibular foramen is on the medial surface.

The mylohyoid groove extends downward and forward from the mandibular foramen.

BONES OF THE EYE ORBIT, PALATE, AND NOSE

Ethmoid Bone: SINGLE (Fig. 24)

A bone of delicate texture, the ethmoid helps to form the floor of the anterior cranial fossa and enters into the formation of the walls of the orbital and nasal fossa. It is shaped much like a fallen capital letter E, with an extended middle portion. It articulates with the frontal, sphenoid, two nasals, vomer, two lacrimals, two maxillae, two inferior nasal conchae and two palatine bones.

LATERAL VIEW

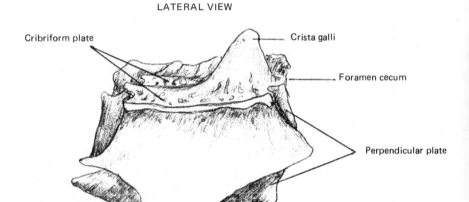

Cribriform plate

Crista galli

Foramen cecum

Perpendicular plate

Fig. 24a. Ethmoid bone

Fig. 24b. Cross section of Ethmoid bone

Cribriform plate—resembling a piece of screen wire, it is a horizontal plate containing many foramina for passage of the olfactory nerves. In the center of the anterior portion is the crista galli.

Crista galli—a thick, triangular plate of bone extending above the cribriform plate. The highest point is in front and its short anterior border joins the frontal bone to form the foramen cecum.

Perpendicular plate—extending below the crista galli and at right angles to the cribriform plate, it forms the upper and posterior third of the nasal septum.

BONES OF SIMILAR SHAPE WHERE CONFUSION MAY ARISE:

None.

INFORMATION OF IMPORTANCE IN IDENTIFICATION:

The cribriform plate and crista galli are very characteristic.

Lacrimal Bone: PAIRED (Fig. 25)

A very thin and delicate bone situated in the anterior part of the medial wall of the eye socket. Immediately posterior to the frontal portion of the maxilla it articulates with the ethmoid, maxilla, frontal and inferior nasal concha.

LATERAL VIEW

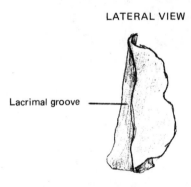

Lacrimal groove

Fig. 25. Left Lacrimal

ANATOMICAL CHARACTERISTICS OF IMPORTANCE IN IDENTIFICATION:

Lacrimal groove—in the anterior portion of the bone is a deep depression. With the maxilla this forms the lacrimal groove that holds the lacrimal sac and duct.

BONES OF SIMILAR SHAPE WHERE CONFUSION MAY ARISE:

None.

Side Identification:

Because of the extremely delicate nature of this bone, one is seldom placed in a position to determine sides. However, the lacrimal groove is always anterior and is better defined on the inferior than on the superior part of the bone.

Palate Bones: PAIRED (Fig. 26)

The two palate bones form the posterior part.of the hard palate and part of the lateral wall of the nasal fossa. They are delicate L-shaped bones. The palatine bones articulate with both maxillae, the sphenoid, the vomer, both inferior nasal conchae, the ethmoid, and its counterpart on the opposite side.

POSTERIOR VIEW

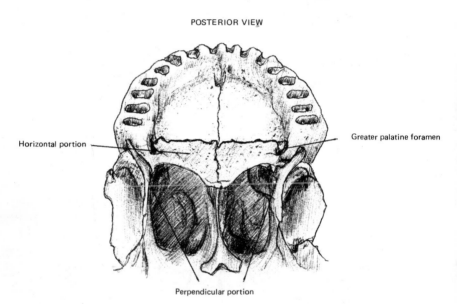

Horizontal portion

Greater palatine foramen

Perpendicular portion

Fig. 26. Palate bones

ANATOMICAL CHARACTERISTICS OF IMPORTANCE IN IDENTIFICATION:

Horizontal portion—forms the posterior part of the bony portion of the roof of the mouth. It articulates anteriorly with the palatine process of the maxilla. Laterally at its junction with the perpendicular portion is the greater palatine foramen.

Perpendicular portion—thinner than the horizontal portion, this portion forms the posterior lateral walls of the nasal fossa.

Greater palatine foramen—a large foramen at the edge of the hard palate and above the second and/or third molars.

BONES OF SIMILAR SHAPE WHERE CONFUSION MAY ARISE:

Sphenoid, because of the delicate structure and L-shape. *Maxilla,* because of the surface of the roof of the mouth.

Side identification:

The horizontal portion is the least fragile and most easily recognized.

Remember that the posterior border is a nonarticular edge.

The greater palatine foramen is always lateral.

Vomer: SINGLE (Fig. 27)

Sometimes called the ploughshare bone, it is a thin, flat bone that lies in the median plane and forms part of the lower and posterior portions of the nasal septum. It articulates with the ethmoid, sphenoid, two palate bones and the two maxillae.

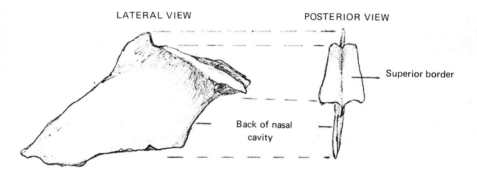

Fig. 27. Vomer

ANATOMICAL CHARACTERISTICS OF IMPORTANCE IN IDENTIFICATION:

Superior border—the thickest part of the bone; articulates with the sphenoid and is expanded laterally into two small wings.

BONES OF SIMILAR SHAPE WHERE CONFUSION MAY ARISE:

Especially the greater and lesser wings of a fragmentary *sphenoid,* and possibly the *nasal.*

INFORMATION OF IMPORTANCE IN IDENTIFICATION:

The expanded superior border is characteristic and offers the best area of identification.

Inferior Nasal Concha: PAIRED (Fig. 28)

Also known as the turbinate bone, it is a slender, very fragile, scroll-like bone attached by its upper margin to the lateral wall of the nasal fossa. The inferior border is free. The superior and middle nasal conchae are a part of the ethmoid bone. The inferior nasal concha articulates with the two maxillae, two lacrimals, two palate bones and the ethmoid.

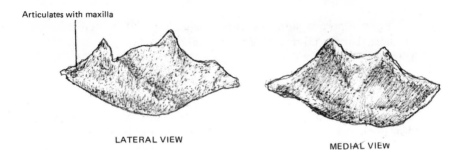

Articulates with maxilla

LATERAL VIEW

MEDIAL VIEW

Fig. 28. Inferior Nasal Concha

Note the rough and uneven texture of the surface. When air is drawn in through the nose this uneven texture causes turbulence and aids in smell.

BONES OF SIMILAR SHAPE WHERE CONFUSION MAY ARISE:

The cerebral surface of the *frontal* bone above the eye orbits has a rough texture, and when fragmentary may cause confusion.

BONES OF THE EAR

Auditory Ossicle: PAIRED (Fig. 29a)

The auditory ossicles consist of three bones in each ear; the malleus, incus and stapes.

MALLEUS (HAMMER): This is the largest and most external of the auditory ossicles and is attached to the tympanic membrane (Fig. 29b). Its club-shaped head articulates with the incus.

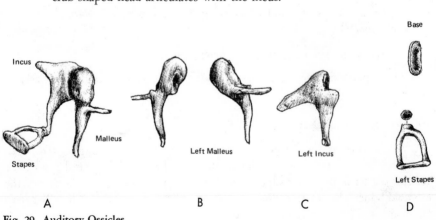

Fig. 29. Auditory Ossicles

INCUS (ANVIL): Situated between the malleus and the stapes, it presents a body and two processes (Fig. 29c). It is the middle of the three ossicles of the ear.

STAPES (STIRRUP): This is the innermost of the ossicles of the ear (Fig. 29d). Shaped somewhat like a stirrup, it articulates by its head with the incus, and its base is inserted into the fenestra ovalis.

MISCELLANEOUS

Hyoid Bone: SINGLE (Fig. 30)

This is a bone that is situated in the anterior part of the neck and does not articulate with any other bones. It supports the tongue and gives attachments to numerous muscles used in speech.

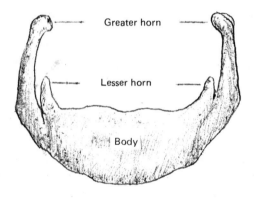

Fig. 30. Hyoid bone (front view)

ANATOMICAL CHARACTERISTICS OF IMPORTANCE IN IDENTIFICATION:

Body—the central portion of the bone that is situated horizontally across the midline of the neck just inferior and posterior to the mandible.

Greater horns—project upward and backward from the sides of the body.

Lesser horns—may be present (and in some individuals are cartilaginous) and are small conical processes that project upward and slightly backward from the anterior surface of the body.

BONES OF SIMILAR SHAPE WHERE CONFUSION MAY ARISE:

The *sphenoid* because of the greater and lesser wings and the *zygomatic* and *temporals* because of the zygomatic arch, which to some resemble the greater horns. In the postcranial skeleton it is often confused with the *vertebrae* (their posterior and lateral projections).

A few minutes studying this bone should result in a fairly comprehensive knowledge of it. Both pairs of horns are slender and project backward and upward from the body.

MEASUREMENTS OF THE SKULL

For the past two decades, following the publication of Washburn's "The New Physical Anthropology" (1951: 298-302), there has been a tendency to discredit the importance of measurements in physical anthropology. I believe that this was not Washburn's intent but instead he was cautioning against the taking of long lists of anthropometric measurements for measurements sake. The simple measuring of every angle, protuberance and bone does not aid in a more accurate description of a skeleton or a skeletal population.

The researcher must first know what type of information he is seeking. Anthropometry then offers only one means of obtaining the desired data. Howells (1969:451-458) has an excellent article on "Criteria for Selection of Osteometric Dimensions" and reference to it is strongly recommended for anyone planning to measure a single skull or bone or a whole population.

Anthropometric measurements are not outdated but rather are techniques for obtaining precise data which can then be used in either univariate statistics or in the more recent and rapidly expanding field of multivariate techniques. The use of anthropometric measurements in multivariate analyses can be found in Howells (1969a:451-458; 1969b:311-314; 1970a:1-3; 1970b:269-272), Giles (1964: 129-135; 1970a:59-61; 1970b:99-109), and Giles and Elliot (1962:147-157; 1963:53-68).

This manual was written for the purpose of training students and for use by professional anthropologists as well as the student who needs to use the techniques of physical anthropology to secure accurate data to solve related problems. The information on measurements which follow are for the purpose of training the unskilled in the use of instruments and techniques. Measurements either inaccurately taken or taken from the wrong anatomical landmarks are not only useless but are a waste of time for the researcher.

Anthropometric measurements are important. So long as researchers wish to compare skeletal populations, some type of measurements must be used. A knowledge of the proper instruments, techniques and anatomical landmarks will aid the researcher in achieving accurate raw data for further use. This data may be used in conjunction with other techniques in physical anthropology such as blood typing, dermatoglyphics, chemical and microscopic analyses and others to aid us in our understanding of man.

Anthropometry is a technique for the measurement of men, whether living or dead. Anthropometry can be divided into:

Somatometry—measurement of the body of the living and of cadavers
Cephalometry—measurement of the head and face
Osteometry—measurement of the skeleton and its parts
Craniometry—measurement of the skull

Professionals are often called upon to compare not only the bones and anatomical features, but also the contours and proportions of skulls and long bones. Observations must be translated into objective measurements, and relationships between measurements may suggest the period of time in which the individual lived.

In order to take anthropometric measurements on bones one must learn the proper position of the bones and points or anthropometric landmarks from which measurements are made.

FRANKFORT HORIZONTAL (PLANE): FH is a plane passing through three points of the right and left porion and the left orbitale. The skull is usually oriented in this plane when drawings, photographs, or illustrations are to be made for comparative or illustrative purposes. When the eyes are fixed on the horizon the skull is normally in a position so that all three points used to determine the FH are in the horizontal plane. First proposed at the meeting of the Craniometric Congress held at Munich, Germany in 1877, it was later ratified at the International Congress of Anthropologists in Frankfort, Germany in 1884, hence the name.

Anthropometric points:

There are two types of anthropometric points:

1. Unpaired: A single point which falls on the mid-sagittal plane (on the midline of the body). For example, there is only one point at the tip of the nose.
2. Paired: Two points that are equidistant on either side of the mid-sagittal plane. For example, if there is a point at the tip of the right ear, there must be a corresponding point at the same location on the left ear.

Measurements are taken between points or anatomical landmarks (Table 11, 12); thus the distance between two points can be given (i.e., the length of the skull is 190 millimeters).

Indices express the ratio of the width to the length of an object. The *Cranial Index* expresses the ratio of the width of the skull to its length.

There have been literally hundreds of anthropometric landmarks defined, and these can be found by consulting the selected references at the end of this chapter. It is my intent here to give only those in most common use. Landmarks are listed in many forms—alphabetical, by section of body or skull, etc. I have found it very helpful for the student to start at the chin and work around the skull so that the landmarks will be in sequence.

The following section on Craniometry has been arranged with the problems of the measurer in mind. The measurements needed to calculate an index are given (numbered) and are followed immediately by the index (lettered) which can be calculated from the measurements. Classification of the index units are also given.

Landmarks of the Skull

Most of these refer to precise points upon the skull surface, either external or internal (Figs. 31-33; Tables 11, 12).

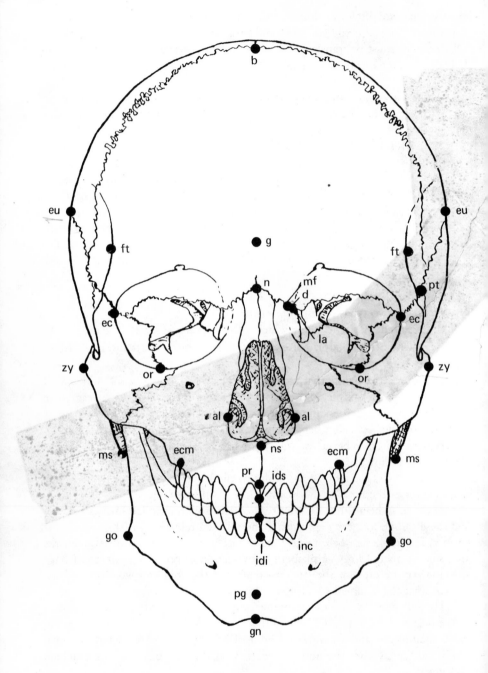

Fig. 31. Selected anthropometric landmarks of the skull (frontal view)

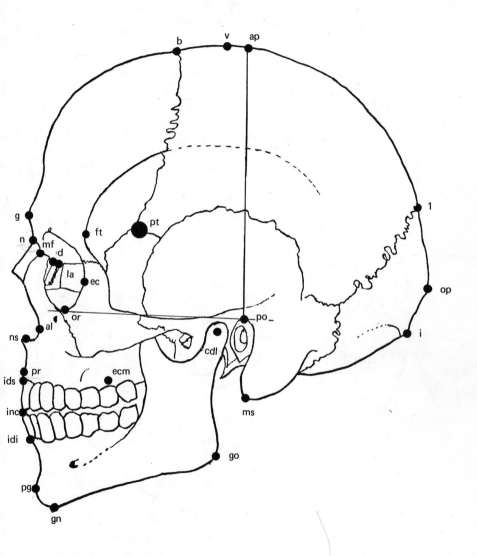

Fig. 32. Selected anthropometric landmarks of the skull (lateral view)

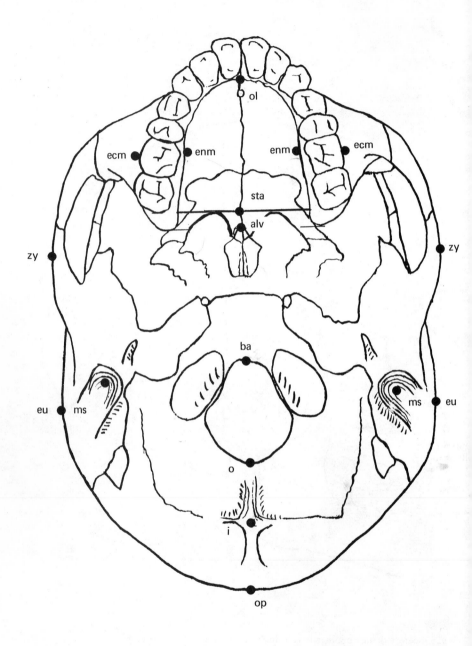

Fig. 33. Selected anthropometric landmarks of the skull (basilar view)

TABLE 11. CRANIOMETRIC POINTS ON THE M.S.P. (MID-SAGITTAL PLANE) (UNPAIRED POINTS)

Facial Skeleton

Gnathion (gn)—The lowest median point on the lower border of the mandible.

Pogonion (pg)—The most anterior point on the chin in the midline.

Infradentale (idi)—(The lower alveolar point): the apex of the septum between the lower central incisors.

Incision (inc)—The incisal level of the upper central incisors.

Alveolare (ids) (infradentale superius)—(The upper alveolar point): the apex of the septum between the upper central incisors. The lowest landmark for the measurement of facial height.

Prosthion (pr)—(Prealveolar point): has often been confused with alveolare. Prosthion is the most anterior point on the upper alveolar process in the midline.

Nasospinale (ns)—Draw a line connecting the lower margins of the right and left nasal apertures. Ns is where this line is intersected by the MSP (mid-sagittal plane). NS is the lowest landmark for the measurement of nasal height.

Nasion (n)—The midline point of intersection of the internasal suture with the naso-frontal suture; the point where the two nasal bones and the frontal bones come together. Nasion is the uppermost landmark for the measure of facial height.

Braincase

Glabella (g)—The most forward projecting point of the forehead in the midline at the level of the supra-orbital ridges and above the naso-frontal suture.

Bregma (b)—The intersection of the coronal and sagittal sutures, in the midline.

Vertex (v)—The highest point in the mid-sagittal contour, as seen from norma lateralis (the lateral view in the Frankfort Horizontal), when the cranium is in the FH.

Apex (ap)—With skull in the FH, draw a perpendicular line to the FH through porion. The apex is where this perpendicular intersects the mid-sagittal contour.

Lambda (1)—The intersection of the sagittal and lambdoidal sutures in the midline.

Opisthocranion (op)—The most posterior point on the skull not on the external occipital protuberance. It is the posterior end point of maximum cranial length measured from glabella. It is thus not a fixed point but is instrumentally determined.

Inion (i)—At the base of the external occipital protuberance. It is the intersection of the MSP with a line drawn tangent to the uppermost convexity of the right and left superior nuchal line.

Opisthion (o)—The midpoint of the posterior margin of the foramen magnum.

Basion (ba)—The midpoint of the anterior margin of the foramen magnum most distant from the bregma. It is used to measure the height of the skull.

Endobasion (endoba)—The most posterior point of the anterior border of the foramen magnum on the border or contour of the foramen. It is thus placed a bit behind and internal to basion. In this way it gives the maximum basi-nasal and basi-prosthion dimensions. (Not shown on Figures.)

Hard Palate

Alveolon (alv)—A point on the hard palate where a line drawn through the termini of the alveolar ridges crosses the median line.

Staphylion (sta)—The point in the midline of the back of the hard palate (inter-palatal suture) where it is crossed by a line drawn tangent to the curves of the posterior margin of the palate.

Orale (ol)—A point on the hard palate where the line drawn tangent to the curves in the alveolar margin back of the two medial incisor teeth crosses the MSP. It is on the opposite side of the bone from the alveolare.

TABLE 12. CRANIOMETRIC POINTS LATERAL TO THE MSP (PAIRED POINTS)

Euryon (eu)—The two points on the opposite sides of the skull that form the termini of the lines of greatest breadth, i.e., the most widely separated points on the two sides of the skull. The two points are instrumentally determined.

Porion (po)—The uppermost lateral point in the margin of the external auditory meatus. The right and left porion with the left orbitale define the FH.

Mastoidale (ms)—The lowest point on the mastoid process.

Dacryon (d)—On the medial wall of the orbit at the junction of the lacrimo-maxillary suture and the frontal bone.

Lacrimale (la)—The point of intersection of the posterior lacrimal crest with the fronto-lacrimal suture.

Maxillofrontale (mf)—The point of intersection of the anterior lacrimal crest (medial edge of eye orbit), or the crest extended, with the fronto-maxillary suture.

Alare (al)—The instrumentally determined most lateral point on the nasal aperture taken parallel to the nasal height.

Orbitale (or)—The lowest point in the margin of the orbit; one of the points used in defining the FH.

Zygion (zy)—The most lateral point of the zygomatic arch; a point instrumentally determined.

Ectoconchion (ec)—The point where the orbital length line, parallel to the upper border, meets the outer rim. Ectoconchion is the point of maximum breadth on the lateral wall of the eye orbit.

Ectomolare (ecm)—The most lateral point on the outer surface of the alveolar margins, usually opposite the middle of the upper second molar tooth. Used in taking the maxillary breadth.

Pterion (pt)—This is a region, rather than a point, and designates the upper end of the greater wing of the sphenoid, with the bodering bones, frontal, parietal, and temporal.

Endomolare (enm)—The most medial point on the inner surface of the alveolar ridge opposite the middle of the second upper molar tooth; used in taking the palatal breadths.

Condylion laterale (cdl)—The most lateral point on the condyle of the mandible.

Gonion (go)—The midpoint of the angle of the mandible between body and ramus. In practice this is hard to determine in jaws with a rounded angle. Draw a line tangent to the posterior border of the ascending ramus. Draw another tangent to the body of the mandible. Bisecting this angle will give the point gonion on the lateral surface of the mandible.

Frontotemporale (ft)—The most medial point on the incurve of the temporal ridge. The points lie on the frontal bones just above the zygomatico-frontal suture.

Measurements and Indices of the Skull

The measurements of the skull selected for description in this manual are listed in Table 13.

TABLE 13. DESCRIBED CRANIAL MEASUREMENTS

No.	Measurement	Page
1	Maximum cranial length	62
2	Maximum cranial breadth	62
3	Basion-bregma height (maximum height)	62
4	Porion-bregma height	65
5	Basion-porion height	66
6	Auricular height	67
7	Minimum frontal breadth	67
8	Total facial height	67
9	Upper facial height	67
10	Facial width or Bizygomatic breadth	67
11	Nasal height	68
12	Nasal breadth	68
13	Orbital height	69
14	Orbital breadth	69
15	Maxillo-alveolar length	70
16	Maxillo-alveolar breadth	70
17	Palatal length	71
18	Palatal breadth	71
19	Bicondylar breadth	72
20	Bigonial breadth	72
21	Height of ascending ramus	72
22	Minimum breadth of ascending ramus	72
23	Height of mandibular symphysis	72

Indices derived from the measurements and described in the manual are listed in Table 14.

TABLE 14. DESCRIBED CRANIAL INDICES

Letter	Index	Page
A	Cranial Index (CI)	63
B	Cranial Module	64
C	Cranial Length-Height index	64
D	Cranial Breadth-Height index	65
E_1	Mean Height Index	65
E_2	Mean Basion-Height Index	65
F	Mean Porion-Height Index	66
G.	Index Flatness of the Cranial Base	67
H.	Fronto-Parietal Index	67
I.	Total Facial Index	68
J.	Upper Facial Index	68
K.	Nasal Index	69
L.	Orbital Index	69
M.	Maxillo-Alveolar Index	71
N.	Palatal Index	71

Craniometric Dimensions (Record all measurements in millimeters)

On the Cranial Vault (Fig. 34)

1. Maximum length (spreading caliper). From glabella to opisthocranion.
   Place one end of spreading caliper on glabella and support it with your finger. With the other end, locate the most posterior point on the midline (opisthocranion) and record length in millimeters.
2. Maximum breadth (spreading caliper). From euryon to euryon.
   The maximum width or breadth is instrumentally determined as both ends of the spreading caliper are moved back and forth on the sides of the skull, above the supramastoid crest until the maximum width is located. Be careful of skulls with warped temporal bones. Sometimes the temporals have spread out and width should not be taken from these.
3. Basion-bregma height (maximum height) (spreading caliper). From basion to bregma.
   Place skull on side or back and hold one end of caliper on basion. Place other end on bregma. If bregma is depressed (the sutures are much below the exterior surface of the vault) take the reading from the surface and not in the depression.

These three measurements can be used to calculate five indices for comparison of size and shape.

A. Cranial Index (CI) (The term Cephalic Index refers to measurements of the living): a numerical device for expressing the ratio of the breadth of the skull to the length (in percent). A skull whose breadth is the same as its length would have a CI of 100.00.

$$\text{Cranial Index} = \frac{\text{Maximum Cranial breadth x 100}}{\text{Maximum Cranial length}}$$

$$\boxed{\text{Example}} \quad CI = \frac{147 \times 100}{172} = \frac{14700.00}{172} = 85.47$$

Range:
Dolichocrany—X-74.99—narrow or long headed
Mesocrany—75.00-79.99—average or medium
Brachycrany—80.00-84.99—broad or round headed
Hyperbrachycrany—85.00-X—very broad

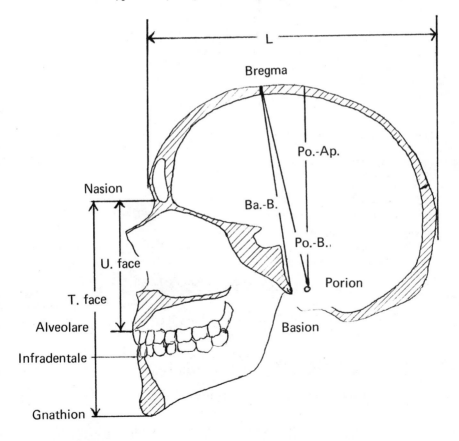

Fig. 34. Cranial measurements

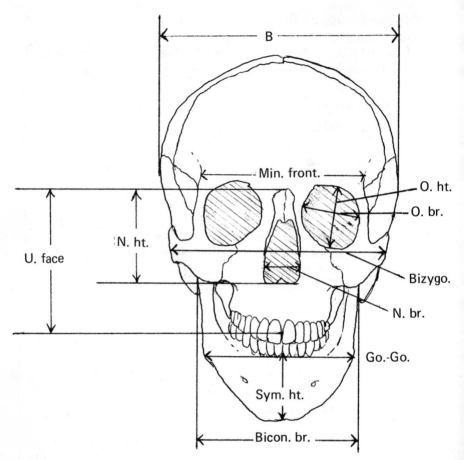

Fig. 34. Cranial measurements (cont.)

In general, when skulls are studied from a evolutionary point of view, man's skull has tended to become more brachycranic. Early fossil men usually have dolichocranic or long heads.

B. Cranial Module: provides a rough numerical value for the size of the skull.

$$\text{Cranial module} = \frac{\text{Length} + \text{breadth} + \text{height}}{3}$$

C. Cranial Length-Height Index: expresses the ratio of height to length of a skull (in percent). See Index E.)

$$\text{Length-Height Index} = \frac{\text{Basion-bregma height x 100}}{\text{Maximum length}}$$

Range: (Martin, 1928, gives these divisions)
Chamaecrany—X-69.99—low skull
Orthocrany—70.00-74.99—average or medium
Hypsicrany—75.00-X—high skull

D. Cranial Breadth-Height Index: expresses the ratio of height to breadth of a skull (in percent). (See Index E.)

$$\text{Breadth-Height Index} = \frac{\text{Basion-bregma height x 100}}{\text{Maximum breadth}}$$

Range: (Martin, 1928, gives Broca's division as)
Tapeinocrany—X-91.99—low skull
Metriocrany—92.00-97.99—average or medium
Acrocrany—98.00-X—high skull

Stewart (1940: 25-30) proposed the use of a Mean Height Index to replace both the Length-Height and Breadth Height Indices because it seemed to be a more sensitive indicator of anatomical change in the skull. Stewart included this measurement in his 1952 edition of Hrdlicka's *Practical Anthropometry* and suggested the formula and range.

$$\text{E}_1 \ \text{Mean Height Index} = \frac{\text{Basion-bregma height x 100}}{\text{Mean of cranial length + breadth}}$$

$$\boxed{\text{Example}} = \frac{135 \text{ x } 100}{\dfrac{172 + 147}{2}} = \frac{13500.00}{159.5} = 84.01$$

Range: (Stewart's, 1952, divisions—see Hrdlicka, 1952)
Low—X-80.49
Medium—80.50-83.49
High—83.50-X

E_2. Mean Basion-Height Index: this is exactly the same as the Mean Height Index but has been given a new name and new classificatory divisions have been suggested.

$$\text{Mean Basion-Height Index} = \frac{\text{Basion-bregma height x 100}}{\dfrac{\text{Cranial length + breadth}}{2}}$$

New Range: (Stewart 1965: 363)
Low—X-78.99
Medium—79.00-85.99
High—86.00-X

In a more recent article Stewart (1965: 359-365) draws attention to the fact that the base of the skull is often damaged or missing and thus eliminates much valuable data. He suggests that the following indices be used.

In order to calculate the second of Stewart's indices (Mean Porion-Height Index) the following measurement is necessary.

4. Porion-Bregma Height (Western Reserve (Todd) Head Spanner). From porion to bregma.

Place the two ear rods in the external auditory meatus on the right and left porions and the end of the calibrated bar on bregma. Read porion-bregma height directly from the calibrated bar.

F. Mean Porion-Height Index: compares the height of the skull from porion with the mean of the length plus the breadth. This measurement can be obtained when the face and base of the skull are missing and only the cranial vault remains.

$$\text{Mean Porion-Height Index} = \frac{\text{Porion-bregma x 100}}{\dfrac{\text{Cranial length + breadth}}{2}}$$

Range: (Stewart 1965: 364)

Low—X-66.99

Medium—67.00-71.99

High—72.00-X

NOTE: Stewart states that when more comparative data are available, the above tentative classification may have to be revised.

Another effective index for classification of skulls according to cranial height has been proposed by Neumann (1942: 178-191). In addition to the measurements already given, the following measurement is necessary to calculate the Index of Flatness of the Cranial Base.

5. Basion-porion height (coordinate caliper) (Fig. 35). From basion to porion.

Place the ends of the sliding caliper on the right and left porion. Move the coordinate attachment until it is over basion and read basion-porion height from calibrated bar when tip of bar is placed on basion.

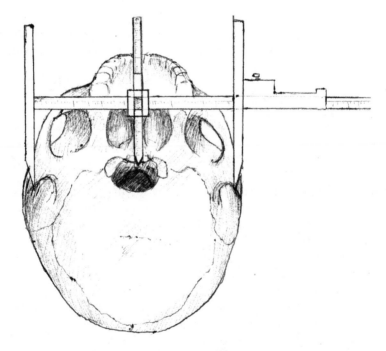

Fig. 35. Use of Coordinate caliper to take Basion-porion height.

G. Index of Flatness of the Cranial Base: Neumann (1942) says that cranial height has two components, basion-porion and porion-bregma heights. He feels that crania with low vaults are the result mainly of flattening of the cranial base, i.e., a short basion-porion distance.

$$\text{Index of Flatness of the Cranial Base} = \frac{\text{Basion-porion height x 100}}{\text{Basion-bregma}}$$

Range. None have been proposed but Neumann sample is:
Low—average around 13.70 (Aleut, Apache)
High—average around 18.40 (Ohio series)

6. Auricular height (Western Reserve (Todd) Head Spanner). From porion to the apex.

The head spanner has a device for orientation in the Frankfort Horizontal. The two horizontal pieces fit into the ear openings (porion) and when the attachment is placed on the left orbitale, the auricular height is read directly when the calibrated bar is placed at the apex. It should be noted that this measurement depends upon the presence of the face to locate the left orbitale. Porion-bregma height is similar to this measurement and does not require the presence of the face.

7. Minimum frontal breadth (spreading or sliding caliper). From fronto-temporale to frontotemporale).

This is measured on the temporal line and at the point where they are closest together, i.e., minimum distance between the temporal crests on the frontal bone.

H. Fronto-Parietal Index: expresses the relationship between the minimum breadth of the frontal bone and the maximum breadth of the vault.

$$\text{Fronto-Parietal Index} = \frac{\text{Minimum frontal breadth x 100}}{\text{Maximum cranial breadth}}$$

Range:
Stenometopic—X-65.99—narrow
Metriometopic—66.00-69.99—average or medium
Eurymetopic—70.00-X—broad

On the Facial Skeleton

Heights

8. Total facial height (sliding caliper). From nasion to gnathion.

With teeth occluded place fixed limb of caliper at nasion and movable end on gnathion. This gives the height of the complete face.

9. Upper facial height (sliding caliper). From nasion to alveolare.

This gives the height of the face excluding the teeth and mandible. It is used when the mandible is missing.

Widths

10. Facial width or Bizygomatic breadth (spreading or sliding caliper). From zygion to zygion. It is the greatest breadth between the zygomatic arches.

These three measurements can be used to determine the size of the face. Two indices express the overall relationship of height to breadth of the face.

I. Total Facial Index: a numerical expression of the ratio of the height to the breadth of the face including the teeth.

$$\text{Total Facial Index } = \frac{\text{Total facial height x 100}}{\text{Bizygomatic breadth}}$$

Range:

Hypereuryprosopy—X-79.99—very broad face
Euryprosopy—80.00-84.99—broad face
Mesoprosopy—85.00-89.99—average or medium
Leptoprosopy—90.00-94.99—slender or narrow face
Hyperleptoprosopy—95.00-X—very slender or narrow face

J. Upper Facial Index: gives a numerical expression of the height to breadth of the face that does not include the mandible teeth.

$$\text{Upper Facial Index} = \frac{\text{Upper facial height x 100}}{\text{Bizygomatic breadth}}$$

Range: (Martin, 1928, divisions)

Hypereuryeny—X-44.99—very wide or broad face
Euryeny—45.00-49.99—wide or broad face
Meseny—50.00-54.99—average or medium
Lepteny—55.00-59.99—slender or narrow face
Hyperlepteny—60.00-X—very slender or narrow face

The Nose (Fig. 36a)

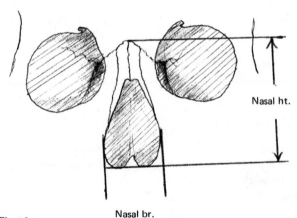

Nasal ht.

Fig 36a.

Nasal br.

11. Nasal height (sliding calipers). From nasion to nasospinale. Place fixed point of caliper at nasion and with movable point obtain the mean of the lowest points of the right and left nasal margins (nasospinale).

12. Nasal breadth (sliding caliper). From alare to alare. The maximum breadth of the nasal cavity. Measured at a right angle to the height.

K. Nasal Index: a numerical method of expressing the relationship of breadth to height of the anterior nasal aperture.

$$\text{Nasal Index} = \frac{\text{Nasal breadth x 100}}{\text{Nasal height}}$$

Range:

Leptorrhiny—X-47.99—narrow nasal aperture
Mesorrhiny—48.00-52.99—average or medium
Platyrrhiny—53.00-X—broad or wide nasal aperture

The Orbits (left is standard for comparison) (Fig. 36b)

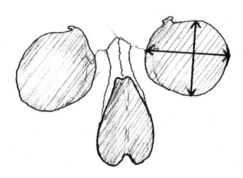

Fig. 36b

13. Orbital height (sliding caliper). The maximum height from the upper to the lower orbital borders perpendicular to the horizontal axis of the orbit and using the middle of the inferior border as a fixed point. Either or both orbits may be measured but the left is standard.

14. Orbital breadth (width) (sliding caliper). From maxillofrontale to ectoconchion. The maximum distance of the orbit from maxillofrontale to the middle of the lateral orbital border (ectoconchion). Measurement can also be taken from dacryon or lacrimale but I prefer maxillofrontale since this is most often present. Since bones of the medial wall of the eye orbit are quite fragile, dacryon and lacrimale are often missing in archaeological specimens. To locate maxillofrontale extend the medial edge of the eye orbit with a pencil line until the line crosses the fronto-maxillary suture.

L. Orbital Index: expresses the relationship of height to breadth.

$$\text{Orbital Index} = \frac{\text{Orbital height x 100}}{\text{Orbital breadth}}$$

Range:

Chamaeconchy—X-82.99—wide orbits
Mesoconchy—83.00-89.99—average or medium
Hypsiconchy—89.00-X—narrow orbits

The Palate (Fig. 37)

External Measurements

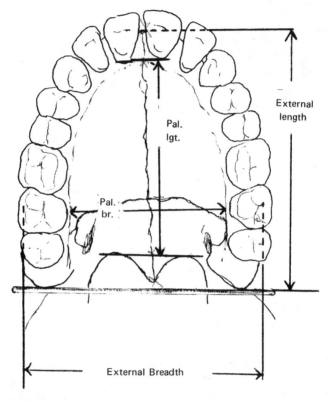

Fig. 37

15. Maxillo-alveolar length (palatal length) (sliding or hinge caliper). From prosthion to alveolon. On a skull with protruding teeth it is difficult to take this measurement with a sliding caliper. Place one end of the caliper on prosthion and the other on a straight wire, knitting needle or wood rod placed across the posterior edges of the alveolar processes (alveolon) of the two sides.

16. Maxillo-alveolar breadth (Palatal breadth) (sliding or hinge caliper). From ectomolare to ectomolare (biectomolare). The distance between the external surfaces of the alveolar border, usually opposite the second molar teeth. If there are any exostoses (bony growths projecting outward) on the border they are to be avoided by placing the ends of the caliper in an unaffected area.

M. Maxillo-alveolar Index: numerical ratio of the external measurements of the palate. Note that this is one of the few indices in which the smaller number is divided into the larger giving an index of over 100.

$$\text{Maxillo-alveolar Index} = \frac{\text{Maxillo-alveolar breadth x 100}}{\text{Maxillo-alveolar length}}$$

Range:
Dolichurany—X-109.99—long or narrow palate
Mesurany—110.00-114.99—average or medium
Brachyurany—115.00-X—broad palate

Internal Measurements

17. Palatal length (sliding caliper). From orale to staphylion. From the median point of a line tangent to the posterior alveolar border of the median incisors (orale) to the median point of a transverse line connecting the most anterior points of the notches in the posterior border of the palate.

18. Palatal breadth (sliding calipers). From endomolare to endomolare (biendomolare). The greatest transverse breadth between the inner limits of the alveolar arch opposite the second molar teeth.

N. Palatal Index: numerical ratio of the internal measurements of the palate.

$$\text{Palatal Index} = \frac{\text{Maximum palatal breadth x 100}}{\text{Maximum palatal length}}$$

Range:
Leptostaphyline—X-79.99—narrow palate
Mesostaphyline—80.00-84.99—average or medium
Brachystaphyline—85.00-X—broad palate

The Mandible (Fig. 38a)

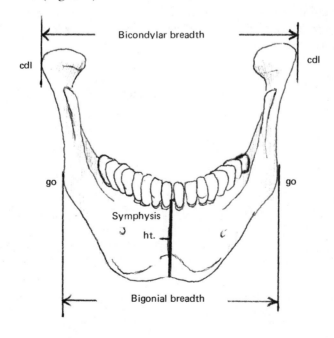

Fig. 38a.

19. Bicondylar breadth (Fig. 38a) (sliding caliper). From condylion to condylion (lateral). The maximum distance between the lateral surfaces of the condyles.
20. Bigonial breadth (Fig. 38a) (sliding caliper). From gonion to gonion. The maximum distance between the external surfaces of the gonial angles.

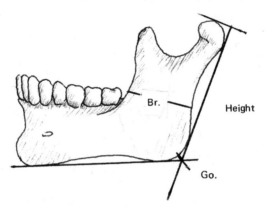

Fig. 38b.

21. Height (length) of ascending ramus (Fig. 38b) (sliding caliper). From gonion to the uppermost part of the condyle. Locate gonion by bisecting the angles formed by prolonged lines drawn on the inferior and posterior borders of the bone, and mark its location. Place fixed end of caliper on top of the condyles and bring movable end to gonion.
22. Minimum breadth of ascending ramus (Fig. 38b) (sliding caliper). Minimum distance between the anterior and posterior borders of the ascending ramus. Can be taken on either right or left but left is standard for comparison.
23. Height of mandibular symphysis (Fig. 38a) (sliding caliper). From gnathion to infradentale. Height in the midline from lowest point (gnathion) to the tip of bone between lower central incisors (infradentale).

Sex Estimation

The skull is probably the second best area of the skeleton to determine sex. Estimation of sex is based upon the generalization that the male is more robust, rugged and muscle-marked than the female. Absolute differences seldom exist and many intermediate forms are found, but distinguishing characteristics are as follows:

I. Face (Fig. 39a)
   1. Supra-orbital ridges are more prominent in males than in females.
   2. Upper edges of the eye orbits are sharp in females, blunt in males.
   3. The palate is larger in males.
   4. Teeth are larger in males.

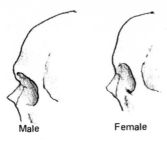

Male   Female

Fig. 39a.

II. Mandible (Fig. 39b)
1. The chin is more square in males and rounded with a point in the mid-line in females.
2. Teeth are larger in males.

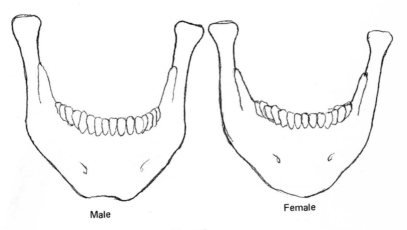

Male   Female

Fig. 39b.

III. Vault (Fig. 39c,d)
1. The female skull is smaller, smoother and more gracile. The female skull retains the childhood characteristics of frontal and parietal bossing into adulthood (Keen 1950).
2. Muscle ridges, especially on the occipital bone, are larger in males (nuchal crests).

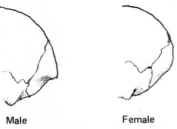

Male   Female

Fig. 39c.

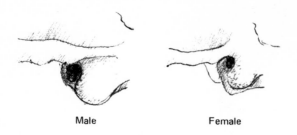

Male   Female

Fig. 39d.

3. The posterior end of the zygomatic process extends as a crest further in males, often much past the external auditory meatus.
4. Mastoid processes are larger in males.
5. Frontal sinuses are larger in males.

III. POSTCRANIAL SKELETON

The postcranial skeleton comprises all of the bones below (after or behind) the skull. Since there are 29 (the hyoid is counted with the skull) bones in the skull there should be 177 bones in the postcranial skeleton to account for the usual 206 bones in the adult human skeleton. Most of these are paired bones, there being a right and a left.

The following general list (Table 15) may help the student to better understand the number of bones in the postcranial skeleton.

TABLE 15. BONES OF THE POSTCRANIAL SKELETON

UNPAIRED BONES

Vertebrae:	Cervical	7	
	Thoracic	12	
	Lumbar	5	
	Sacrum	1	-- usually 5 segments that fuse in adults
	Coccyx	1	-- usually 4 or 5 segments that may fuse in adults
Sternum		1	-- usually 3 segments that may fuse in adults
	Total	27	

PAIRED BONES

Shoulder girdle: (Includes sternum above)	Scapula	2	
	Clavicle	2	
Ribs		24	
Upper extremities:	Humerus	2	
	Radius	2	
	Ulna	2	
	Carpus	16	
	Metacarpus	10	
	Phalanges	28	
	Total		88
Pelvic girdle: (Includes sacrum above)	Hip bone	2	
Lower extremities:	Femur	2	
	Patella	2	
	Tibia	2	
	Fibula	2	
	Tarsus	14	
	Metatarsus	10	
	Phalanges	28	
	Total		62
Total number of postcranial bones		-	177

VERTEBRAL COLUMN

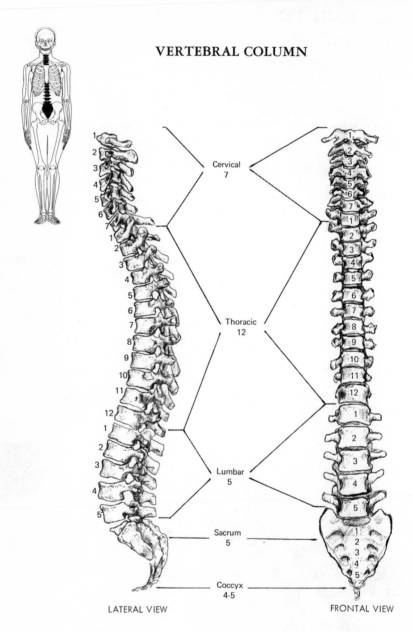

Cervical
7

Thoracic
12

Lumbar
5

Sacrum
5

Coccyx
4-5

LATERAL VIEW

FRONTAL VIEW

Fig. 40.

The Vertebral Column: UNPAIRED—IRREGULAR BONES (Fig. 40)

The vertebral column is usually composed of 33 segments. Sometimes one more or less may be found. The upper 24 are separate and are true or movable vertebrae. Of these, 12 articulate with ribs and 12 do not. The next 5 (sacral) rapidly decrease in size, and in adults are fused into a triangular bone, the sacrum, which is one of the three bones of the pelvis. The next 4, or sometimes 5, are the coccygeal or tail vertebrae (coccyx). From head to tail the number and names of the vertebrae are as follows:

Type	Region	No.
Cervical	Neck	7
Thoracic	Chest	12
Lumbar	Lower back	5
		24
Sacral	Pelvis	5
Coccygeal	"Tail"	4
		9 – 33

Subadult bones:

Ossification of a vertebra takes place in cartilage from three primary centers (that appear from 7-20 weeks intra-uterine) and five secondary centers (that appear about puberty).

At birth each typical vertebra consists of three bony parts, the centrum and each half of the arch (Fig. 41a). Union (synosteosis) of the two halves of the arch takes place posteriorly during the 1st to 3rd year (some say as late as the 7th year) and the arch and body fuse between the 3rd and 7th year (Fig. 41b). The vertebrae attain almost their full size by puberty but epiphyses have not fused.

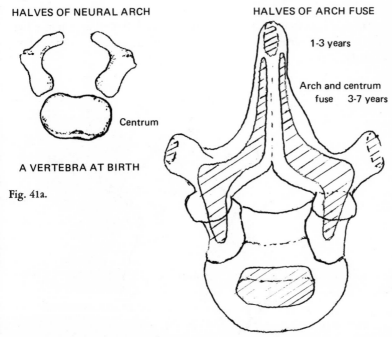

HALVES OF NEURAL ARCH

HALVES OF ARCH FUSE

1-3 years

Arch and centrum
fuse   3-7 years

Centrum

A VERTEBRA AT BIRTH

Fig. 41a.

Fig. 41b.

FUSION OF 3 PRIMARY CENTERS OF A VERTEBRA

With the exception of some of the upper cervical vertebrae, all present at least five epiphyses—upper and lower epiphyseal rings of the centrum (term body refers to both centrum and epiphyses), tips of the spinous and both transverse processes (Fig. 41c). These epiphyses appear about puberty and fuse between 17-25 years.

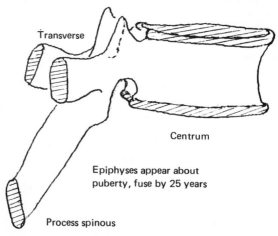

Transverse

Centrum

Epiphyses appear about
puberty, fuse by 25 years

Process spinous

Fig. 41c. Fusion of 5 secondary centers of a vertebra.

In man centers for upper and lower surfaces of body form marginal epiphyseal rings, but are complete plates in other mammals, i.e.—dogs, sheep.

The surfaces of immature centra (Fig. 41d) have a billowed appearance that resemble epiphyseal surfaces elsewhere in the skeleton and especially the symphyseal surface of the pubis.

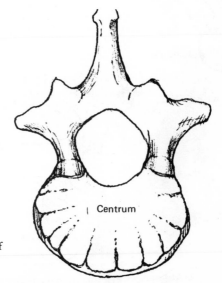

Centrum

Fig. 41d. Billowed surface of
immature vertebra

Adult bones:

The bones of each region have characteristic features that are particular to that segment of the spinal column but every bone in the column has one or more distinguishing features of its own.

MOST VERTEBRAE (FIG. 42) ARE COMPOSED OF FOUR PARTS:

1. Body—weight bearing portion (except for the first cervical vertebra which has no body).
2. Vertebral arch—part that protects the spinal cord.
3. Spinous process and right and left transverse processes—for attachment of muscles.
4. Four articular processes; two superior (above) and two inferior (below)— for articulation with vertebrae above and below.

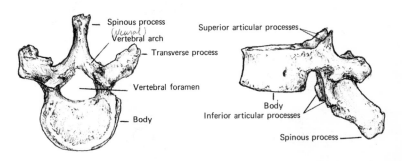

Fig. 42.

In the first 24 vertebrae, the size of the body increases progressing from head to tail to support the increased weight. When called upon to reassemble the vertebrae of a disarticulated column, give first consideration to the relative massiveness of the bodies and later regard articular facets and inclination of articular processes. Above the sacrum all of the body weight is supported on a single column, the vertebral column. Below the pelvis, weight is supported on two columns, the legs.

Variations do occur in the number of vertebrae. Whereas the number of cervical vertebrae is quite constant at 7, additional vertebrae can occur in both the thoracic and lumbar regions. Less common is a missing vertebra in these latter regions.

CERVICAL VERTEBRAE

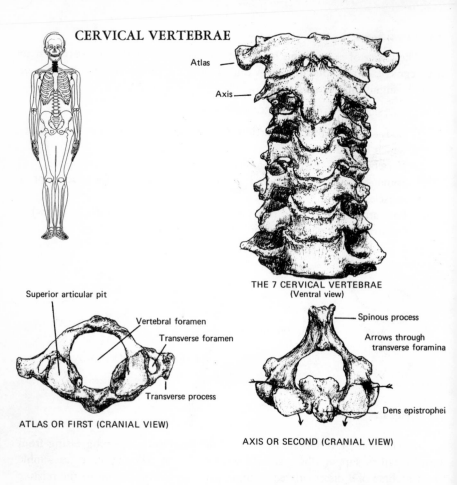

Atlas

Axis

THE 7 CERVICAL VERTEBRAE
(Ventral view)

Superior articular pit

Vertebral foramen

Transverse foramen

Transverse process

ATLAS OR FIRST (CRANIAL VIEW)

Spinous process

Arrows through
transverse foramina

Dens epistrophei

AXIS OR SECOND (CRANIAL VIEW)

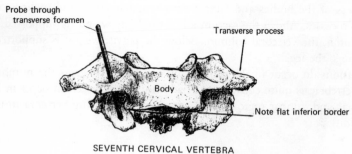

Probe through
transverse foramen

Transverse process

Body

Note flat inferior border

SEVENTH CERVICAL VERTEBRA

Fig. 43.

Cervical Vertebrae: 7 SINGLE—IRREGULAR BONES (Fig. 43)

It is through the cervical vertebrae that the skull articulates with the post-cranial skeleton, and these vertebrae possess a high degree of flexibility. They are the smallest of the movable vertebrae. All mammals, with few exceptions, have seven.

Type	Region	No.
Cervical	Neck	7

DISTINGUISHING FEATURES OF CERVICAL VERTEBRAE:

1. All have transverse foramina. A transverse foramen is found only in cervical vertebrae.
2. Cervical vertebrae are smaller in size when compared with the thoracic and lumbar vertebrae but increase in size from 1 through 7.
3. The ventral (front) portion of the articular surface of the body of Numbers 3, 4, 5, 6, and the upper border of 7 are lower in the center than on the sides.

BONES OF SIMILAR SHAPE WHERE CONFUSION MAY ARISE:

Thoracic and possibly lumbar vertebrae. These are larger and do not have transverse foramina.

OF SPECIAL INTEREST ARE:

Number 1, the *Atlas*
This is the only true vertebra that has no body or central mass of bone, and it has no spinous process. It is merely a large ring upon which the skull rests.

Number 2, the *Axis* or Epistropheus
The second vertebra is easily identified because of the Dens epistropheus which forms a pivot on which the atlas, carrying the head, rotates. The Dens, sometimes known as the odontoid process (the displaced body of the atlas) has an articular facet on its ventral (front) surface for articulation with the atlas.

Number 7
A transitional vertebra situated at the junction of the cervical and thoracic regions, this presents characteristics of both regions:
1. It has the largest body of any cervical vertebra.
2. It has a flat or straight inferior (bottom) edge of the body.

THORACIC VERTEBRAE

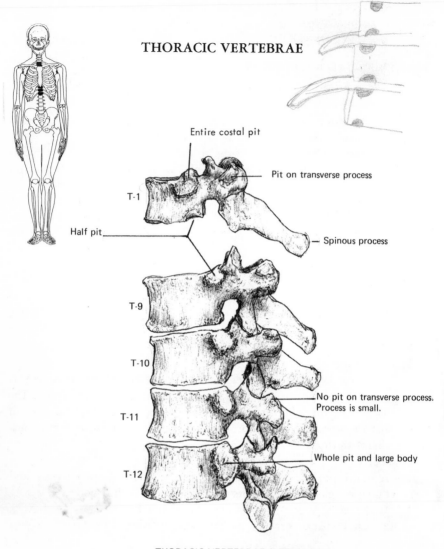

Entire costal pit

Pit on transverse process

T-1

Half pit

Spinous process

T-9

T-10

No pit on transverse process.
Process is small.

T-11

Whole pit and large body

T-12

THORACIC VERTEBRAE (LEFT VIEW)

Fig. 44.

Thoracic Vertebrae: 12 SINGLE—IRREGULAR BONES (Fig. 44)

The thoracic vertebrae support the (12) ribs and enter into the composition of the thorax.

Type	Region	No.
Thoracic	Chest (Mid-back)	12

1. All have costal pits on the sides of the bodies and on most of the transverse processes for articulation with the ribs. Most of the thoracic vertebrae have costal pits at the borders; one-half at the superior and one-half at the inferior border so placed that each completes, with the adjacent vertebra, a cavity for the head of a rib.
2. None of the thoracic vertebrae have transverse foramina.

BONES OF SIMILAR SHAPE WHERE CONFUSION MAY ARISE:

Lumbar Vertebrae. Check carefully for articular pits (facets) which occur on the body of thoracic vertebrae only, and for the angle of the articular processes, which form a sharp angle in lumbar vertebrae but are parallel in thoracic vertebrae (Fig. 45).

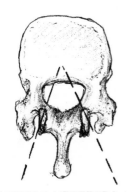 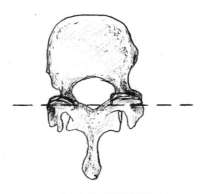

LUMBAR VERTEBRA
(note articular angle)

THORACIC VERTEBRA
(parallel articular surfaces superior and inferior)

Fig. 45a.   Fig. 45b.

OF SPECIAL INTEREST ARE:

Number 1
    This has a whole and a half costal pit.
Number 2 through 9
    All have half pits, on the superior and the inferior body.
Number 10
    This has whole pits on the body and on the transverse process.
Number 11
    This has a whole pit on the body but none on the transverse process.
Number 12
    This looks like No. 11, except that the inferior articular surfaces are not parallel and assume the lumbar pattern.

LUMBAR VERTEBRAE

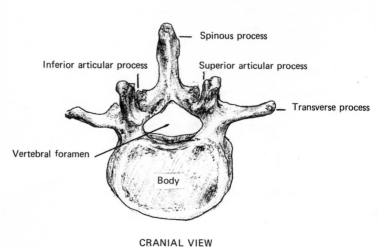

Spinous process

Inferior articular process

Superior articular process

Transverse process

Vertebral foramen

Body

CRANIAL VIEW

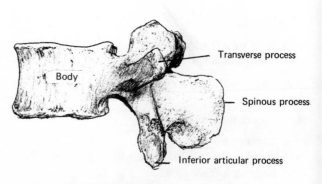

Transverse process

Body

Spinous process

Inferior articular process

LATERAL VIEW

A LUMBAR VERTEBRA

Fig. 46.

Lumbar Vertebrae: 5 SINGLE—IRREGULAR BONES (Fig. 46)

The lumbar vertebrae are the largest of the presacral vertebrae and they support the weight of the body above the pelvis.

Type	Region	No.
Lumbar	Lower Back	5

DISTINGUISHING FEATURES OF THE LUMBAR VERTEBRAE:

1. No transverse foramen.
2. No costal pits for articulation of ribs..
3. Largest of the movable vertebrae.
4. Transverse spines slant upward.
5. Spinous processes are larger and are more horizontal than in thoracic vertebrae.
6. The superior and inferior articular facets are U-shaped (Fig. 45a) (for greater support) and are different from the parallel type articulation of the cervical and thoracic vertebrae. (Fig. 45b).

BONES OF SIMILAR SHAPE WHERE CONFUSION MAY ARISE:

Thoracic vertebrae. Note that there are no articular facets on the body for ribs, and the angles of the articular processes are different from the thoracic vertebrae (Fig. 45).

SACRUM AND COCCYX

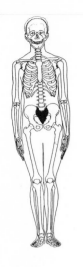

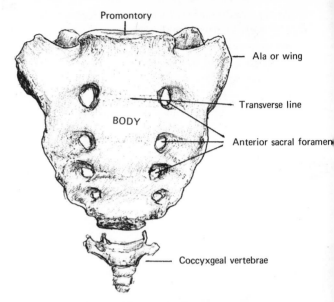

Promontory

Ala or wing

Transverse line

BODY

Anterior sacral foramen

Coccyxgeal vertebrae

ANTERIOR OR PELVIC VIEW

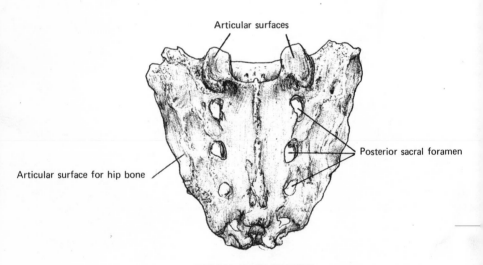

Articular surfaces

Posterior sacral foramen

Articular surface for hip bone

DORSAL OR BACK VIEW

Fig. 47.   THE SACRUM AND COCCYX

Sacrum: 5 SINGLE—IRREGULAR BONES (Fig. 47)

and

Coccyx: 4 SINGLE—IRREGULAR BONES (Fig. 47)

Subadult bones:

The sacrum ossifies from 35 centers with each of the five sacral segments having the typical vertebra's three primary centers plus 2 for the costal elements (Fig. 48). Segments 4 and 5 have no costal centers. The costal elements begin to fuse with each other about puberty.

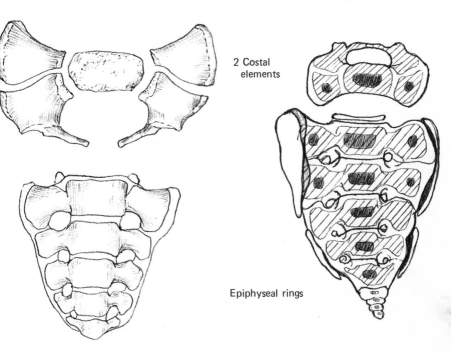

3 Primary centers

2 Costal elements

Epiphyseal rings

SACRUM AND COCCYX AT AGE 4   **Fig. 48.**   OSSIFICATION OF SACRUM

Epiphyseal rings for bodies appear around puberty with the rings and bodies fusing from below upward in the 18-25 year range. Late in life the coccyx often fuses with the sacrum.

Discussing final epiphyseal closure, McKern and Stewart (1957: 97) state:

> As early as 18 years but any time between 18 and 25 years, the epiphyseal cap begins to unite to the billowed surface of the medial end of the clavicle. Union begins at the approximate center of the face and spreads to the superior margin where it may progress either anteriorly or posteriorly. From 25 to 30, the majority of cases are undergoing terminal union. The last site of union is located, in the form of a fissure, along the inferior border. With the obliteration of these fissures (at age 31), the epiphysis is completely united.

Adult bones:

The sacrum is usually composed of five separate bones that unite into a single triangular shaped unit in adulthood. It is a large, curved, wedge-shaped bone that forms the base of the vertebral column and firmly connects with the two hip bones at the sacro-iliac joint. The sacrum rapidly decreases in size from the first vertebra (which is the largest) to the fifth (which is rudimentary).

The sides of the upper three vertebrae of the sacrum form a large articular surface which joins with the iliac portion of the hip bones. The ala or wing is the large portion of bone between the body of the sacrum and the articular surfaces. The sacrum articulates with the fifth lumbar vertebra superiorly at the promontory, with the iliac portion of the innominate (hip) bones laterally, and the coccyx inferiorly.

The anterior and posterior sacral foramina transmit nerves, arteries and veins.

Type	Region	No.
Sacral	Pelvis	5
Coccygeal	"Tail"	4-5

BONES OF SIMILAR SHAPE WHERE CONFUSION MAY ARISE:

When the sacrum is fragmentary it is possible to confuse it with the *lumbar vertebrae,* especially the fourth and fifth. Remember that the lumbar vertebrae do not have alae.

Measurements of the sacrum (Fig. 49a):

Maximum anterior height: (sliding calipers).

From the middle of the sacral promontory to the middle of the antero-inferior border of the last sacral vertebra (usually the fifth). For comparative purposes use only sacra with 5 segments (A-B).

Maximum anterior breadth: (sliding calipers).

The greatest distance between the wings (lateral masses) of the first sacral vertebra (C-D).

Sacral index:

$$\frac{\text{Sacral anterior breadth} \times 100}{\text{Sacral anterior height}}$$

Racial indices of the sacral index
(Wilder 1920: 118)

	Males	Females
Negroes	91.4 (33)	103.6 (18)
Egyptians	94.3 (7)	99.1 (2)
Andamanese	94.8 (22)	103.4 (35)
Australians	100.2 (14)	110.0 (13)
Japanese	101.5 (37)	107.1 (30)
Europeans	102.9 (63)	112.4 (43)

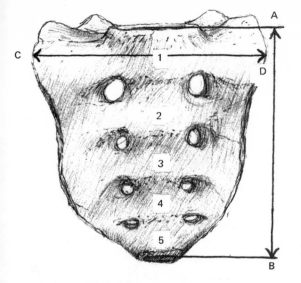

Fig. 49a.

Sex determination (Fig. 49b):

1. The sacrum is generally more curved in males and flatter in females.
2. In some cases the width of the body of the sacrum to the ala is greater in males. Anderson (1962: 142) states that in females the width of the first sacral body (articular area) is equal in width to each ala.

OBSERVATIONS:

Record the number of segments in the sacrum, four to six (see numbers on Fig. 49a). Note also a sacralized (partly fused) fifth lumbar or a first sacral vertebra that is partly free (lumbarized). In older individuals, note arthritic lipping.

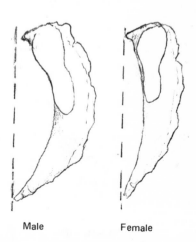

Male   Female

Fig. 49b.   SACRUM, SIDE VIEW

STERNUM

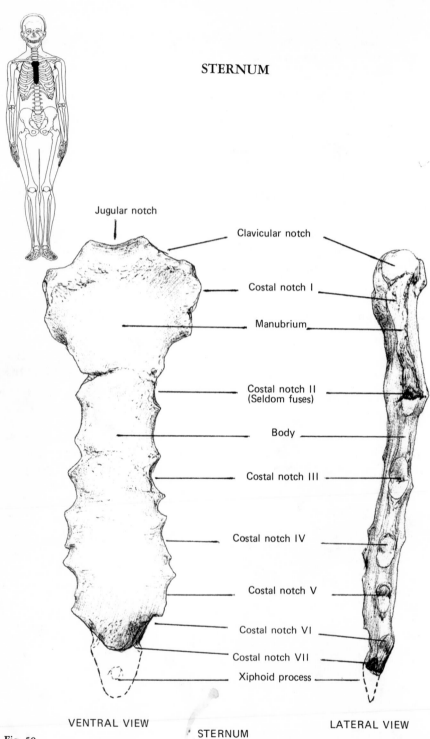

Jugular notch

Clavicular notch

Costal notch I

Manubrium

Costal notch II
(Seldom fuses)

Body

Costal notch III

Costal notch IV

Costal notch V

Costal notch VI

Costal notch VII

Xiphoid process

VENTRAL VIEW

STERNUM

LATERAL VIEW

Fig. 50.

90

STERNUM: Single—Flat Bone (Fig. 50)

The sternum or breast bone is a flat plate of bone situated in the ventral (front) wall of the thorax. It is similar to a broad sword, and has three parts:

1. Manubrium—handle
2. Corpus sterni—body or blade
3. Xiphoid process—tip

In young individuals it is composed of six segments:

1st—remains separate and forms the manubrium.

2nd-5th—fuse to form the body.

6th—remains separate and forms the tip.

Manubrium—the broadest and thickest part of the bone. The suprasternal or jugular notch can easily be felt in the midline at the superior end of the sternum.

Body—longer, narrower, and thinner than the manubrium. The lines of union of the four pieces making up the body can be seen in the adult bone.

Xiphoid process—thin and the least developed of the parts of the sternum. It is cartilaginous in early life, partly ossified in adults, and in old age tends to become ossified throughout and fused with the body.

Note that the sternum is slightly concave on the dorsal (back) surface.

BONES OF SIMILAR SHAPE WHERE CONFUSION MAY ARISE:

None.

Sex determination:

In many cases the body of the sternum in males is more than twice the length of the manubrium. In females the body is less than twice the length of the manubrium.

SCAPULA

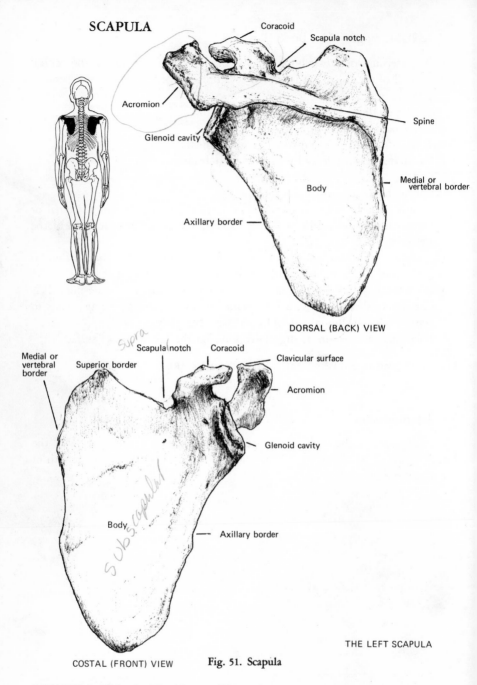

Coracoid

Scapula notch

Acromion

Glenoid cavity

Spine

Body

Medial or vertebral border

Axillary border

DORSAL (BACK) VIEW

Medial or vertebral border

Superior border

supra

Scapula notch

Coracoid

Clavicular surface

Acromion

Glenoid cavity

subscapular

Body

Axillary border

THE LEFT SCAPULA

COSTAL (FRONT) VIEW

Fig. 51. Scapula

SCAPULA: Paired—Flat Bones (Fig. 51)

Subadult bones:

The ossification of the scapula occurs from two primary centers (body of scapula and coracoid) and seven secondary centers (Fig. 52).

The coracoid fuses with the scapula beginning about the 15th year along a line that includes the upper part of the glenoid cavity and is usually completed by the 18th year.

The scapula usually presents six epiphyses:

Location of Epiphyses	Age of Union
2 for the coracoid process	15–18 years
1 for the glenoid cavity	15–18
1 for the acromion	16–22
1 for the inferior angle	17–22
1 for the medial (vertebral) border	17–23

There appears to be great variability in the union of the epiphyses of the scapula as can be seen in the table under Age Determination.

Adult bones:

The scapula is a large flat bone, and it is triangular in shape. It is situated on the dorsal (back) side of the thorax between the level of the 2nd and 7th ribs. From the bone there are two projecting processes: (1) the coracoid process (for muscle attachments) and (2) the spine, which terminates laterally in the acromion (which articulates with the clavicle). The function of the scapula is to

Fig. 52. Ossification centers of scapula.

give attachments to muscles and to form the socket of the shoulder joint where it articulates with the humerus.

The greater part of the scapula consists of a thin triangular plate of bone known as the body, from which the coracoid extends anteriorly and the spine posteriorly.

The medial or vertebral border is the longest of the three borders and, particularly in its middle third, can present a convex, straight or concave pattern. The axillary border is the thickest. It extends from the lower margin of the glenoid cavity to the inferior angle of the bone. The superior border is the shortest of the three borders and terminates laterally at the scapula notch. The scapula notch shows great variability, it may be almost absent, may present a notch which is either shallow, medium or deep, or may be a complete foramen.

FOUR PARTS OF THE BONE ARE IMPORTANT IN IDENTIFICATION:

1. Body (which is concave when viewed from the anterior or front side; sometimes the anterior side is called the costal surface).
2. Coracoid process (anterior projection).
3. Spine (ending in the acromion or posterior projection).
4. Glenoid cavity (articulator for head of humerus).

BONES OF SIMILAR SHAPE WHERE CONFUSION MAY ARISE:

Hip bones (innominates). Both are flat bones but the rest of the anatomy is not similar. The scapula body is thinner than the hip bones and usually presents thinner and sharper borders. In some cases the fragmentary scapula may be confused with the flat bones of the *cranium*. It should be remembered that cranial bones have an inner and outer table of hard compact bone and a diploë of loose tissue between these. Diploë is not found in most of the scapula and no suture line occurs in the scapula.

Side identification:

1. Hold spine toward you, concave side away from you; both the coracoid and the acromial processes, which should be superior (on top), point to the side bone is from.
2. With spine toward you and the coracoid superior, the vertebral border is always toward the midline (along the thoracic vertebrae) and therefore is opposite the side the scapula comes from, i.e. when held in this position, if the vertebral border is on the right side of the bone, the bone is from the left side of the body and vice versa.

The scapula in man, and to a lesser extent in some of the primates, is expanded anterio-posteriorly far in excess of that of most other mammals. The infra-spinous process is the main part that has lengthened in man.

Measurements of the scapula (Fig. 53):

Maximum length (total height): (sliding caliper or osteometric board). The maximum straight line distance (A-B) from the superior to the inferior border.

Maximum breadth: (sliding caliper). From the middle of the dorsal border of the glenoid fossa to the end of the spinal axis on the vertebral border (C-D).

Length of spine: (sliding caliper). From the end of the spinal axis on the vertebral border (same point as above) to the most distal point on the acromion process (D-E).

Length of supra-spinous line: (sliding caliper). From the end of the spinal axis on the vertebral border (same point as above) to the top of the anterior angle (A-D).

Length of infra-spinous line: (sliding caliper). From the end of the spinal axis on the vertebral border to the tip of the posterior angle (D-B).

$$\text{Scapula index:} \quad \frac{\text{Maximum breadth} \times 100}{\text{Maximum length}}$$

OBSERVATIONS:

I. Form of the vertebral border (Fig. 54).
    The vertebral border may be: convex, straight, concave or any combination of these.

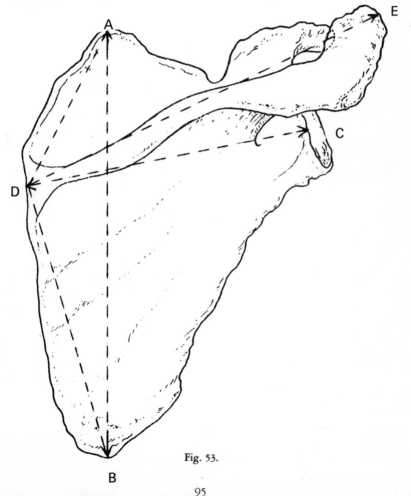

Fig. 53.

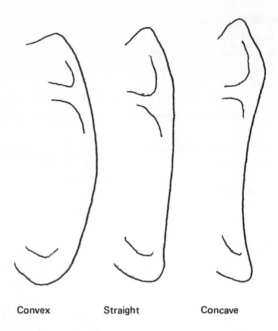

| Convex | Straight | Concave |

Fig. 54.   VERTEBRAL BORDERS

The greatest dimension of the scapula of most mammals (horse, cat, etc.) is in the same direction as the spine; the breadth is at right angles to this (Fig. 55).

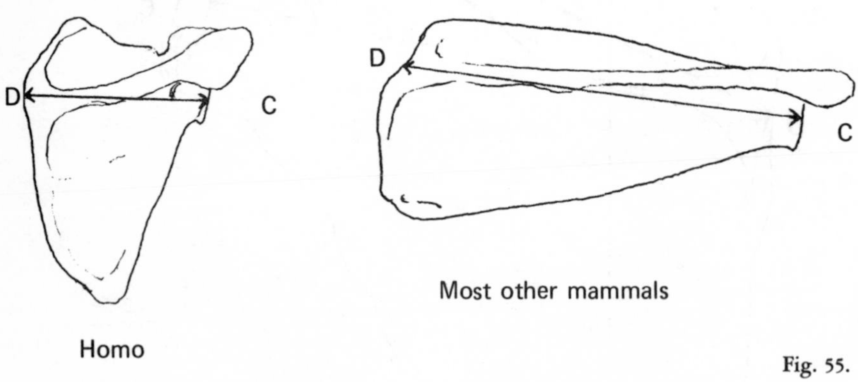

Most other mammals

Homo

Fig. 55.

II. Form of scapula notch (Fig. 56).

The notch formed by the superior border and the coracoid process may be absent, shallow, medium deep or may be a foramen.

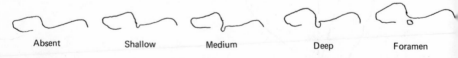

| Absent | Shallow | Medium | Deep | Foramen |

Fig. 56.   FORM OF SCAPULA NOTCH

Age estimation:

Ossification of the scapula and the appearance and union of the epiphyses is given under the subadult section.

The following table (16), taken mainly from Stevenson (1924:54) with information from more recent authors, summarizes the ages assigned to union of the epiphyses.

In the older age range, Graves (1922:27) has noted the appearance of Atrophic spots. "Atrophic spots may be defined as localized, discrete or coalescing areas of bone atrophy" (Graves 1922:27). But it should be noted that *not* every translucent area in the body of the scapula is an atrophic spot, no matter what the degree of translucency. There must also be a localized alteration in the vascularity and structure of the bone in addition to translucency.

Atrophic spots vary in size from 3 mm. to 10 mm. or more, are discrete, and exhibit irregular patterns in their early stages. Later they become ovoid or circular and coalesce to involve the greater portion of the body in some senile bones. In such bones the body becomes very thin, and oval defects varying in size from 2 to 5 mm. may occur in the extremely thin portions. One must differentiate between true and "pseudo-defects" (which are often found in scapulae following body fracture). True and Pseudo-defects ". . . are differentiated from atrophic spots by their larger size, thicker margins and by the absence of alterations in vascularity and bone structure" (Graves 1922:29).

Graves (1922) found that atrophic spots were either absent, small or infrequent in scapulae from individuals under 45 years of age, whereas the scapulae from individuals over 45 years usually showed them in greater number, size and degree.

Sex estimation:

Krogman (1962:205-206) has summarized the data on sex determination from the scapula and his two tables (17, 18) are reproduced here.

TABLE 16. AGES ASSIGNED BY VARIOUS AUTHORITIES FOR UNION OF EPIPHYSES*

Authority Epiphysis	Bryce 1911	Dixon 1912	Dwight 1911	Gegan- baur 1892	Henle 1871	Krause 1909	Lewis 1918	Poirier 1911	Terry 1921	Testut 1921	Thompson 1921
Coaracoid	Puberty	17	14-15	16-18	14-15	16-18	15	20-25	15	14-16	15-17
Acromion	22-25	22-25	18-19	--	16-17	19-21	--	20-25	20	17-18	25
Inf. Angle	25	22-24	20	--	20	21-22	25	--	25	22-25	20-25
Vert. Border	25	22-24	20	--	21-22	21-22	25	25-28	25	20-24	20-25

Authority Epiphysis	Flecker (1942)	Graves (1922)	Hrdlicka (1942c)	Johnson (1961)	McKern and Stewart (1957)	Steven- son (1924)	Terry and Trotter (1953)	Earliest	Latest	Difference (Years)
Coaracoid	--	--	--	--	Pre-17	15	18-25	14	25	11
Acromion	17	--	--	18-19.5	Pre-17-23	19	20	16	25	9
Inf. angle	--	--	20	--	Pre-17-23	19-22	25	17	25	8
Vert. Border	--	22	20	--	Pre-17-23	19-22	25	17	28	11

*Stevenson, AJPA 1924 Vol. 7:54.

TABLE 17.

SCAPULAE: MAIN EARLIER DATA
(PRINCIPALLY AFTER VALLOIS, ARRANGED ON BASIS OF SCAPULAR INDEX IN MALES BY A.H)*

Group	Specimens** M F†	Male			Female			Author
		Height Total	Breadth	Scapular Index	Height Total	Breadth	Scapular Index	
Fuegian	35–28	16.02	9.90	61.8	14.33	9.22	64.3	Vallois
Eskimo	4–4	15.70	9.72	61.9	(15.87)	(9.85)	62.0	Vallois
Finn	72–14	16.55	10.25	62.4	14.80	9.32	63.9	Kajava
New Caledonia	10–5	14.83	9.60	63.6	12.86	8.94	69.5	Sarasin
Europ. White	146–102	16.76	10.65	63.7	13.55	9.05	66.8	Livon
Old Feruv. Indian	55–39	15.83	10.17	64.2	13.78	9.17	66.5	Hrdlička
Fuegian	7–2	15.38	9.88	64.3	14.20	9.40	66.1	Garson
N. W. Indian	10–14	16.52	10.48	64.3	14.07	9.37	66.1	Dorsey
Portuguese	37–20	15.92	10.21	64.4	13.62	9.04	66.5	Correa
Fuegian	4–6			64.8			65.7	Martin
French	78–68	15.92	10.37	65.2	14.11	9.28	65.9	Vallois
U. S. White	70–44	16.40	10.70	65.3	14.40	9.60	66.7	Hrdlička
Mex. Indian	9–12	15.80	10.40	65.5	13.75	9.75	70.7	Hrdlička
Egyptian	6–9			65.9			68.0	Warren
Egyptian	11–6	15.78	10.42	66.5	13.0	9.31	68.6	Vallois
Afr. Black	58–15	15.23	11.19	66.6	13.46	9.01	68.2	Vallois
Amer. Negro	46–18	16.25	10.90	66.8	14.20	9.25	65.0	Hrdlička
So. Mongol	20–4			66.9			(65.1)	Vallois
So. Utah Indian	18–10	15.10	10.15	67.4	13.70	9.70	70.6	Hrdlička
Pecos Indian	79–24	14.74	10.11	68.3	13.42	9.67	73.5	Hooton
Melanesian	20–11			68.6			69.1	Vallois
Melanesian	10–12	14.90	10.29	69.1	13.42	9.20	68.6	Sarasin
Lenape Indian	4–9	15.20	10.60	69.5	13.90	9.90	70.7	Hrdlička
Pima and Pueblo	5–5	15.50	11.05	71.0	13.80	9.95	72.0	Hrdlička
Negrillo	4–6	13.15	10.03	77.1	12.10	8.93	73.8	Vallois

*From Hrdlicka, '42, p. 381.

**Number of specimens not equal for all the measurements, but nearly so.

†First figure denotes male, second female specimens.

TABLE 18.

SCAPULAR DIMENSIONS AND INDICES: RANGES OF VARIATION*

		Height, Total	Height, Infra-spinous	Breadth (Broca's)	Indexes Scapular Total	Infra-spinous
			Male			
All Whites	Range	13.7–19.0	9.8–14.7	8.6–12.4	53.8–85.4	68.1–111.1
(1200)	Means	16.04	12.08	10.49	65.0	86.9
	R.A.**	32.8	40.6	36.2	48.6	49.5
N. A. Indian	(229)	13.0–18.4	9.9–14.8	8.9–12.0	57.3–75.9	66.2–101.3
		15.36	11.69	10.115	65.86	86.52
		35.2	41.9	30.6	28.2	40.6
Alaskan	(239)	13.1–18.4	10.0–15.0	8.7–12.2	54.2–72.5	67.4–94.6
Eskimo		16.22	12.77	10.12	62.4	79.2
		32.7	39.2	34.6	27.7	34.3
Amer. Negro	(126)	14.1–18.7	9.9–14.3	9.0–12.4	58.9–76.9	76.8–111.1
		15.98	11.66	10.66	66.7	91.4
		28.8	36.9	31.9	27.0	37.5
			Female			
All Whites	(457)	11.7–16.8	8.5–13.0	8.1–11.3	55.6–84.7	71.4–116.7
		14.19	10.67	9.39	66.3	88.1
		35.9	42.2	34.1	43.9	51.4
Indian	(179)	11.4–16.4	8.4–13.0	8.3–10.9	58.4–86.8	72.3–114.3
		13.73	10.535	9.615	70.0	91.245
		36.4	43.7	27.0	40.6	46.0
Alaskan	(197)	12.3–17.1	9.2–14.2	8.0–10.3	56.7–76.6	65.0–98.0
Eskimo		14.10	11.06	9.25	65.6	83.6
		34.0	45.2	24.9	30.3	39.5
Amer. Negro	(46)	12.6–16.1	8.8–12.4	8.7–10.6	57.7–76.3	75.6–112.8
		14.17	10.23	9.51	67.2	93.0
		24.7	35.2	20.0	27.7	40.0

*From Hrdlicka, '42, p. 399.
**R.A. = Range/Average Index.

Tables 17 and 18 courtesy of Charles C. Thomas, Publisher, Springfield, Ill.

CLAVICLE

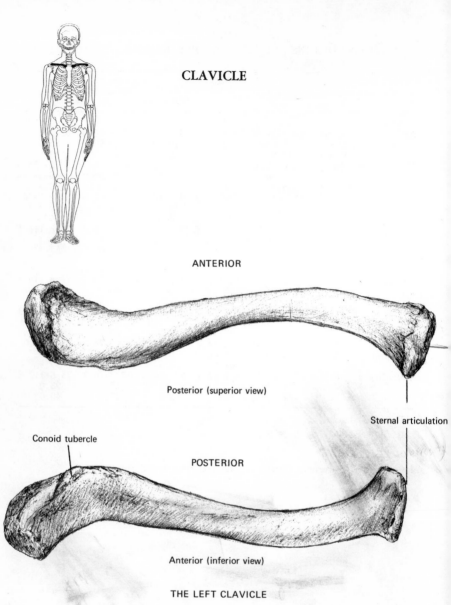

ANTERIOR

Posterior (superior view)

Conoid tubercle

Sternal articulation

POSTERIOR

Anterior (inferior view)

THE LEFT CLAVICLE

Fig. 57.

CLAVICLE: Paired—Long Bone (Figs. 57, 58)

Subadult bone:

The clavicle is the first of all bones of the body to ossify, usually beginning about the fifth week (Fig. 59). A secondary center (epiphysis) appears at the sternal end between 12 and 21 years and is the last of the epiphyses of the body to unite, in most individuals by age 25.

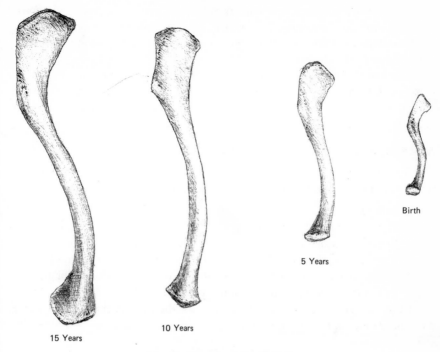

Fig. 58. Growth of the Clavicle.

15 Years

10 Years

5 Years

Birth

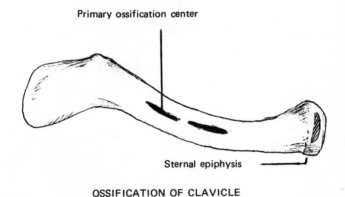

Primary ossification center

Sternal epiphysis

OSSIFICATION OF CLAVICLE

Fig. 59.

Adult bone

The clavicle, or collar bone, is a long bone with a shaft and two ends. It is situated immediately above the first rib, and extends from the upper border of the manubrium (sternum) laterally and backwards to the acromion of the scapula. Its function is to act as a strut or prop to the shoulder, thereby holding the scapula and upper limb laterally, backward and slightly upward.

101

Animals that use their forelimbs merely for support or locomotion (horse, cow, dog, buffalo) have either no clavicles or only rudimentary ones. Animals that use their forelimbs for flying, climbing, grasping or burrowing (primates, rodents, bats) have clavicles.

The loss of the clavicle results in "lengthening" of the limb in a functional sense, *i.e.* instead of swinging from the glenoid the forelimb swings from the muscular attachment between scapula and thorax. Such modification can result in a longer stride (ground distance/degree of movement).

BONES OF SIMILAR SHAPE WHERE CONFUSION MAY ARISE:

The end can be confused with the acromial process of the scapula. The acromial process of the scapula has no conoid tubercle.

Side identification:

1. Shoulder or distal end is flattened.
2. Sternal or medial end is rounded.
3. Conoid tubercle is on the inferior (down) surface near the shoulder end and is always posterior. The conoid tubercle is always inferior (down) and in the back when in anatomical position.

Illustrated (Fig. 60a) is the inferior view of the left and right clavicles with the conoid tubercle facing the viewer.

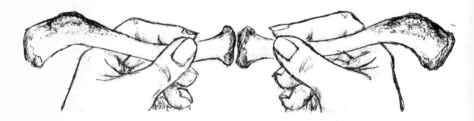

Fig. 60a.

Measurements of the clavicle (Fig. 60b):

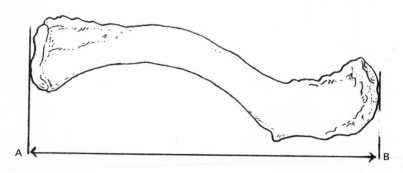

Fig. 60b.

Maximum length: (osteometric board or a sliding caliper). One end of the bone is placed against the stationary end of the board and the movable end of the board is brought into contact with the opposite end of the clavicle. The clavicle is moved from side to side and up and down until the maximum length is obtained (A-B).

Circumference at middle of bone: (metal tape or strip of graph paper). This measurement is taken at the middle of the shaft.

Claviculo-humeral index: Useful as an indicator of the relative development of the thorax.

Claviculo-humeral index:

$$\frac{\text{Maximum length of clavicle x 100}}{\text{Maximum length of humerus (from same side as clavicle)}}$$

Robustness (length: circumference) index: Useful as an indicator of sex.

Robustness (length: circumference) index:

$$\frac{\text{Midclavicular circumference x 100}}{\text{Maximum length of clavicle}}$$

Age determination:

According to McKern and Stewart (1957:91-92) the medial clavicular epiphysis begins to unite in the 17th or 18th year. They found unattached epiphyses as late as the 22nd year but no cases of complete union were found before age 23. They state in summary:

> As early as 18 years but any time between 18 and 25 years, the epiphyseal cap begins to unite to the billowed surface of the medial end of the clavicle. Union begins at the approximate center of the face and spreads to the superior margin where it may progress either anteriorly or posteriorly. From 25 to 30, the majority of cases are undergoing terminal union. The last site of union is located, in the form of a fissure, along the inferior border. With the obliteration of these fissures (at age 31), the epiphysis is completely united.

Sex determination:

The accuracy of determining sex of an individual from the clavicle has met with varying degrees of success—none of them high.

Thieme (1957:72-81) uses clavicle length as one of a series of eight measurements to determine sex in the negro skeleton. Although this single measurement is not too accurate in determining sex, Thieme reports the following data for an American Negro sample.

Measurement	Sex	Number	Mean mm.	Standard Deviation	Standard Error of Mean	Critical Ratio
Clavicle length	M	98	158.24	10.06	1.158	13.90
	F	100	140.28	7.99	0.800	

Jit and Singh (1966:1-21) have studied the sex of adult clavicles from India and report the following:

...there is no such single character which can determined the sex of all clavicles. Depending on the length alone, the sex can be decided in 8 per cent of male and 14 per cent of female right clavicles and in 20 per cent of male and 12 per cent of female bones if the left clavicle is considered. Thus, about 80 to 92 per cent of male clavicles cannot be distinguished by this character. The weight of the clavicle, which can distinguish 24 per cent male bones if the right clavicle is available and 35 per cent if the left is available, is a better guide than the length for the male cases. However, for finding the female clavicles, weight does not help in more than 2 per cent in case of right clavicles and its value is zero per cent for evaluating sex from the left clavicles. The circumference at the middle of the clavicle is of greatest significance because it can distinguish 72 per cent male cases if the right bone is available. However, by left clavicle it can sort out only 48 per cent cases. On the other hand, its utility in the female clavicles is rather poor, being only 10 and 5 per cent in case of right and left bones, respectively. The robustness index has not been found to be of much use in sexing the clavicles, its value being 4 to 8 per cent in male cases and nil in the females. Thus, for positively declaring the bone to be a definitely female one, length alone is more useful than any other single character because by this method 12 to 14 per cent female clavicles can be easily sorted out, whereas by other individual characters the value varies from zero to 10 per cent only. However, by working out linear combinations of different characters a large percentage of female clavicles can definitely be found out, i.e. 21 to 35 per cent from the left and right clavicles, respectively. (Jit and Singh (1966:20).

RIBS

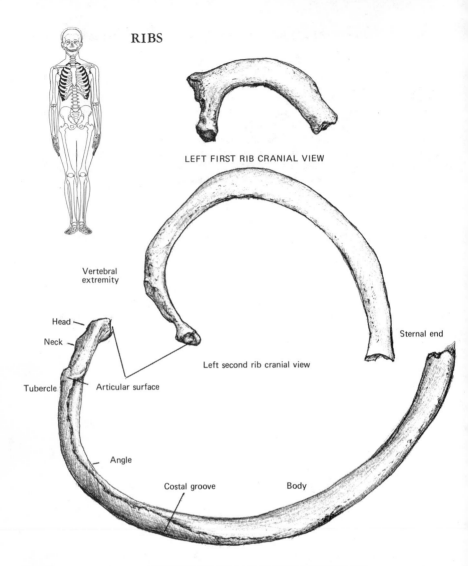

LEFT FIRST RIB CRANIAL VIEW

Vertebral
extremity

Head

Neck

Sternal end

Tubercle   Articular surface

Left second rib cranial view

Angle

Costal groove   Body

Fig. 61.   LEFT SEVENTH RIB CAUDAL VIEW (FROM BELOW)

RIBS: 12 Paired—Flat Bones (Fig. 61)

Subadult bones:

Around the eighth week of intra-uterine life the ribs begin to ossify from a center near the angle of each rib. Ossification progresses rapidly and by the end of the fourth month reaches as far as the costal cartilage (Fig. 62).

Secondary centers for the head and the articular part of the tubercle appear about puberty and fuse between 18 and 24. McKern and Stewart state that "ossification begins in the upper and lower ribs and slowly progresses toward the middle. Thus, the last ribs to become fully united are ribs 4 to 9" (1957:160).

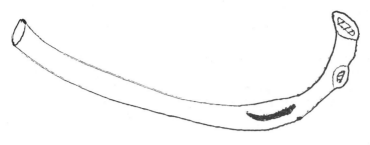

OSSIFICATION OF A TYPICAL RIB

Fig. 62.

Adult bones:

The ribs, 12 on each side, constitute a series of paired, narrow, flattened bones which articulate posteriorly with the vertebral column. The upper seven pairs articulate directly with the sides of the sternum through cartilage and are termed true ribs. The remaining five pairs are classified as false ribs and are of two kinds:

1. The 8th, 9th and 10th ribs have cartilage which connects ventrally (in the front) with the cartilage of the ribs above.
2. On the 11th and 12th, the ventral extremities are free, and are tipped with cartilage.

The classification of the ribs is as follows:

| | Name | |
Rib number	Common	Anatomical
First 7 ribs	true	vertebrosternal
8th, 9th and 10th	false	vertebrochondral
11th and 12th	floating	vertebral

Note that the ribs increase in length from the first through the seventh, and decrease from the eighth through the twelfth. The seventh rib is regarded as the most typical and presents:

1. A vertebral extremity which includes
   a. Head
   b. Neck
   c. Tubercle
2. A shaft or body.
3. A sternal extremity.

The cranial (external or top surface of a rib is convex and usually fairly smooth. The caudal (internal or bottom) surface is concave and contains the costal groove on the lower edge. The cranial edge of the rib is blunt whereas the caudal edge is sharp (with the costal groove on the inside).

First rib: the broadest, most curved, and usually the shortest of all the ribs. The head usually has only one articular facet.

Second rib: longer than the first, strongly curved, but looks more like the ribs below.

BONES OF SIMILAR SHAPE WHERE CONFUSION MAY ARISE:

It is difficult to tell the exact number of the rib without a comparative skeleton.

Side identification:

1. The head is always dorsal (toward the back) and the tubercle is always caudal (down). This is enough to orient the rib, but additional observations to enforce the above for proper orientation are that the thick edge of the rib should be up (cranial) and the costal groove and the sharp inferior (caudal) edge should be down.
2. To tell the sides of the first rib, lay the rib on a table; the head will point down (rest on the table) when the rib is oriented as it is in the body. With the head down the groove for the subclavian vein will be on top.

HUMERUS

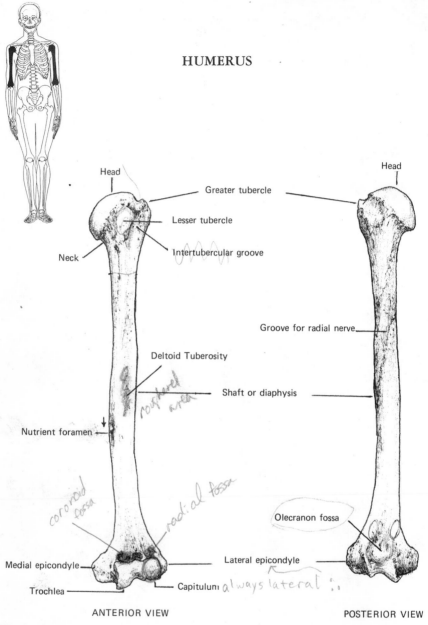

Head

Greater tubercle

Lesser tubercle

Intertubercular groove

Neck

Groove for radial nerve

Deltoid Tuberosity

postered area

Shaft or diaphysis

Nutrient foramen

coronoid fossa

radial fossa

Olecranon fossa

Medial epicondyle

Lateral epicondyle

Trochlea

Capitulum *always lateral :.*

Head

ANTERIOR VIEW

POSTERIOR VIEW

Fig. 63.

LEFT HUMERUS

HUMERUS: Paired—Long Bone (Fig. 63, 64)

Subadult bones:

The humerus ossifies from one primary center (the shaft or diaphysis) and seven secondary centers, three in the proximal or head end and four in the distal end

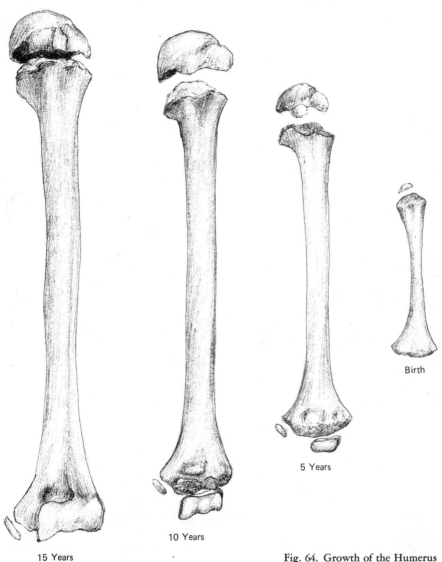

15 Years   10 Years   5 Years   Birth

Fig. 64. Growth of the Humerus

(Fig. 65). The three epiphyses of the proximal end coalesce about the sixth year and fuse with the shaft about the 20th year.

Adult bones:

The humerus is the largest and longest bone in the arm. It articulates proximally with the scapula (at the shoulder) and distally with the radius and ulna (at the elbow). It is divided into a shaft and two extremities:

Fig. 65.   OSSIFICATION OF HUMERUS
After (Lockhard et al 1959:142)

1. Shaft.
2. Proximal extremity:
   a. Head
   b. Neck
   c. Two tubercles, greater and lesser
3. Distal extremity:
   a. Two epicondyles, medial and lateral
   b. Articular sufaces for radius (capitulum) and ulna (trochlea)

The intertubercular groove extends down the lateral side of the bone and lodges the long tendon of the biceps. On the medial side of the bone and in its center one-third is the nutrient foramen. Note that the nutrient foramen (or nutrient canal), through which the nutrient artery enters the bone, is directed toward the distal end of the bone (Fig. 66a). The inclination of the nutrient foramen in the long bones is important in the identification of shaft fragments. If you sit in a squatting position and flex your arms so that your fists are next to the shoulders, the nutrient foramina enters the long bones of the upper and lower extremities in a direction away from the skull.

BONES OF SIMILAR SHAPE WHERE CONFUSION MAY ARISE:

None.

The humerus is smaller than the femur and tibia and larger than the radius, ulna and fibula.

Side Identification:

1. When held (in approximate anatomical position Fig. 66b) with the head toward you and the anterior surface up:

   a. The head is always on the medial or inside and is opposite the side the bone comes from. The medial epicondyle is the same.

   b. The intertubercular groove is anterior, on the lateral side of the bone and is on the same side the bone is from. The capitulum and lateral epicondyle are the same.

2. When you have only the shaft (both ends missing), locate the nutrient foramen and determine its inclination (always toward the distal end). Holding the distal end of the fragment, turn the shaft over so that the nutrient foramen is on the side of the bone opposite from you. The radial groove will be inclined down the bone toward the side the bone comes from (Fig. 66c).

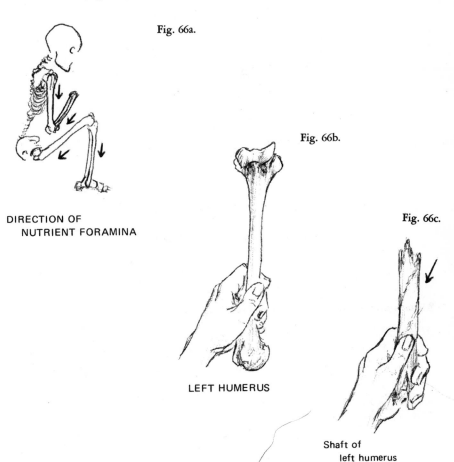

Fig. 66a.

**DIRECTION OF
NUTRIENT FORAMINA**

Fig. 66b.

Fig. 66c.

LEFT HUMERUS

Shaft of
left humerus

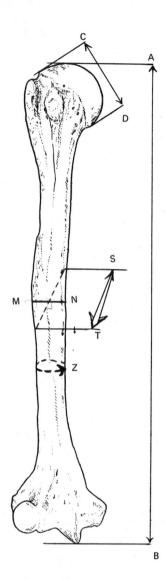

Fig. 67.

Measurements of the humerus (Fig. 67):

Maximum length: (osteometric board)

Place the head against the fixed vertical of the board and adjust the movable upright to the distal end. Raise the bone slightly and move it up and down as well as from side to side until the maximum length is obtained (A-B).

Maximum diameter mid-shaft: (sliding caliper)

Locate the mid-point of the shaft on the osteometric board and mark the bone with a pencil. Measure the maximum diameter; it will be in an anterio-medial direction (M-N).

Minimum diameter mid-shaft: (sliding caliper)

Taken at a right angle to the previous measurements (S-T); it is the minimum diameter of the mid-shaft.

Maximum diameter of head: (sliding caliper)

Taken from a point on the edge of the articular surface of the bone across to the opposite side. The bone is rotated until the maximum distance is obtained (C-D). This measurement is used as an indicator of the sex of the individual.

Least circumference of the shaft: (graduated steel tape)

Taken at about the second third (Z), distal to the deltoid tuberosity. It is usually about a centimeter distal to the nutrient foramen.

Robusticity index: This expresses the relative size of the shaft.

$$\text{Robusticity index} = \frac{\text{Least circumference of shaft x 100}}{\text{Maximum length of humerus}}$$

Radio-humeral index: This expresses the relative length of the forearm to the upper arm.

$$\text{Radio-humeral index} = \frac{\text{Maximum length of radius x 100}}{\text{Maximum length of humerus}}$$

OBSERVATIONS:

Presence of septal apertures (supra-condyloid foramen).

At the distal end of many humeri just above the trochlea there may be a foramen extending through into the olecranon fossa. These foramina, called septal apertures by Hrdlicka (1932:431-450) can appear as: pinpoint, small, medium, large, or they may be absent (Fig. 68). Hrdlicka stated that they occur more frequently in females.

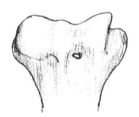 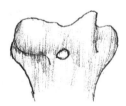 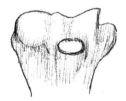

| Small | Medium | Large |

Fig. 68.

Age estimation:

Johnston (1962:249-254) has published on the growth of long bones from the infants and children of the Indian Knoll, Kentucky skeletal population. The relation of age to the length of the subadult bone is as follows where bone length is in millimeters (Table 19).

TABLE 19. AGE ESTIMATION FROM HUMERUS AFTER JOHNSTON'S TABLE 2 (1962:251)

Estimated age in years	Humerus		
	n	Mean	σ
Fetal	9	56.78	6.26
NB–0.5	71	67.66	5.94
0.5–1.5	42	93.14	12.11
1.5–2.5	7	113.57	5.66
2.5–3.5	11	125.64	6.86
3.5–4.5	9	136.78	5.56
4.5–5.5	6	154.67	5.42

McKern and Stewart (1957:44) state that the epiphyses unite as follows:
Distal epiphysis: Completely united by 17-18.
Head of humerus: Completely united by 24.
Medial epicondyle: Completely united by 19.

The medial epicondyle unites from below upward leaving a small notch at the superior end as the last side of union (Fig. 69).

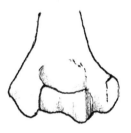

Fig. 69.

Schranz (1959:273-277) has studied radiographically age changes from the internal structure of the humerus based upon 250 macerated humeri. Reported below are his findings beginning with age 23 (he began at age 15).

23-25 years. The development of the metaphysis is accomplished. The internal structure of the epiphysis is no longer quite radial, that of the diaphysis is ogival. The medullary cavity is far from the collum chirurgicum.

26-30 years. The radial arrangement of the internal structure of the epiphysis is fading. The internal structure of the diaphysis is ogival. The medullary cavity has not yet reached the collum chirurgicum.

31-40 years. The internal structure of the epiphysis has lost its earlier characteristic appearance. The internal structure of the diaphysis is more columniform. The most superior parts of the medullary cavity may approach the collum chirurgicum.

41-50 years. The columniform structure of the diaphysis is discontinuous. The cone of the medullary cavity has reached the collum chirurgicum. Between the cone and the epiphyseal line lacunae may be in evidence.

51-60 years. Pea-sized lacunae appear in the tuberculum majus.

61-74 years. The outer surface of the bone is rough and the cortex is thin. The diaphyseal structure lacks characteristic features. The medullary cavity has reached the epiphyseal line. Bean-sized, or even larger, lacunae are present in the tuberculum majus. X-ray pictures of the caput show increased transparency.

Over 75 years. The external surface of the bone is rough. The tuberculin majus has lost its prominence. The cortex is thin. Very little spongy tissue remains in the medullary cavity. In the gross specimen the epiphysis is very fragile. X-ray pictures of the caput show an increased transparency.

In evaluating the above age changes it is important to know whether the bones are male or female because the changes occur at different times in the two sexes. *The difference, which favors the female, amounts to two years at puberty, 5 years at maturity and 7-10 years in senium.* (Schranz 1959: 275-276)

Sex estimation:

The humerus seems to be a poor bone for determining sex. One of the most obvious sex differences in the long bones is that typical male bones are longer and more massive than typical female bones, an expression of sexual dimorphism. However, a number of researchers have studied the humerus and presented the following data in relation to sex determination.

Diameter of the head:

In 1904-05 Dwight (see also Krogman 1962:144) published on the diameter in millimeters of the humeral head:

	Vertical	Transverse
Male	48.76	44.66
Female	42.67	36.98
Difference	6.09	5.68

Humerus length:

Thieme (1957:73) used humerus length and epicondylar width of the humerus in his study of the sex of Negro skeletons. Although neither measurement proved to be a highly accurate indicator he does publish the following data (Table 20).

According to Trotter (1934:214) a septal aperture occurs 3.7 times more frequently in females than in males.

The supratrochlear foramen is one of the better group racial traits in the humerus, and its variation between the sexes also makes it a useful sex indicator. Trotter (1934:217) found the septal aperture in 4.2% of Caucasian and 12.8% of American Negroes. For these same racial groups, Hrdlicka (1932:445) found 6.9% and 18.4% respectively, and 29.6% for American Indians. According to Akabroi (1934:399), the overall average for Japanese is about 17%.

TABLE 20. (AFTER THIEME 1957:73)

Negro

	Sex	No.	Mean (mm)	S.D.	Std. error of mean	Critical ratio (t)
Humerus length	M	98	338·98	18·55	1·874	12·51
	F	100	305·89	18·66	1·866	
Epicondylar width of humerus	M	98	63·89	3·59	0·363	14·50
	F	100	56·76	3·32	0·332	

The female tends to have ossification centers appearing earlier and complete union with the shaft terminating sooner than in the male. Studies by Garn, Rohmann, and Blumenthal found that in cases where roentgenograms are involved, the relationship between the capitulum of the radius and the medial epicondyle of the humerus can be used to determine sex in young specimens. There is a major sexual dimorphism with over 70% discriminatory efficiency in their particular ossification order. They found that the ossification order is capitulum radii-medial epicondyle in at least 75% of the males; 70% of the females had the opposite order (Garn, Rohmann, and Blumenthal 1966:106).

Stature calculations:

The estimation of stature from long bones has been attempted by numerous authors. However, with the humerus, radius and ulna Trotter and Gleser said "it can be stated as a general rule that in no case should lengths of upper limb bones be used in the estimation of stature unless no lower limb bone is available" (1958:120).

Stature formula for humerus - male

White

2.89 Humerus + 78.10 ± 4.57

Negro

2.88 Humerus + 75.48 ± 4.23

Mongoloid

2.68 Humerus + 83.19 ± 4.16

Mexican

2.92 Humerus + 73.94 ± 4.24

Length of humerus must be in centimeters.

Example - for humerus of white male that is 41.0 cm., (410 mm.).

2.89 (41.0) + 78.10 ± 4.57
118.49 + 78.10 ± 4.57

196.59 cm. = mean
Range   196.59 - 4.57 = 192.02 cm. Low
          196.59 + 4.57 = 201.16 cm. High

Only those formulas given by Trotter and Gleser (1952, 1958) will be repeated here (Table 21).

TABLE 21. FORMULAE FOR STATURE ESTIMATION
AFTER TROTTER AND GLESER TABLE 18 (1952:495)

WHITE FEMALES		NEGRO FEMALES	
$3.36 \text{ Hum} + 57.97$	± 4.45	$3.08 \text{ Hum} + 64.67$	± 4.25
$4.74 \text{ Rad} + 54.93$	± 4.24	$2.75 \text{ Rad} + 94.51$	± 5.05
$4.27 \text{ Ulna} + 57.76$	± 4.30	$3.31 \text{ Ulna} + 75.38$	± 4.83
$2.47 \text{ Fem}_m + 54.10$	± 3.72	$2.28 \text{ Fem}_m + 59.76$	± 3.41
$2.90 \text{ Tib}_m + 61.53$	± 3.66	$2.45 \text{ Tib}_m + 72.65$	± 3.70
$2.93 \text{ Fib} + 59.61$	± 3.57	$2.49 \text{ Fib} + 70.90$	± 3.80
$1.39(\text{Fem}_m + \text{Tib}_m) + 53.20$	± 3.55	$1.26(\text{Fem}_m + \text{Tib}_m) + 59.72$	± 3.28
$1.48 \text{ Fem}_m + 1.28 \text{ Tib}_m + 53.07$	± 3.55	$1.53 \text{ Fem}_m + 0.96 \text{ Tib}_m + 58.54$	± 3.23
$1.35 \text{ Hum} + 1.95 \text{ Tib}_m + 52.77$	± 3.67	$1.08 \text{ Hum} + 1.79 \text{ Tib}_m + 62.80$	± 3.58
$0.68 \text{ Hum} + 1.17 \text{ Fem}_m + 1.15 \text{ Tib}_m + 50.12$ [3]	± 3.51	$0.44 \text{ Hum} - 0.20 \text{ Rad} + 1.46 \text{ Fem}_m + 0.86 \text{ Tib}_m + 56.33$	± 3.22

To estimate stature of older individuals subtract .06 (age in years--30) cm

RADIUS

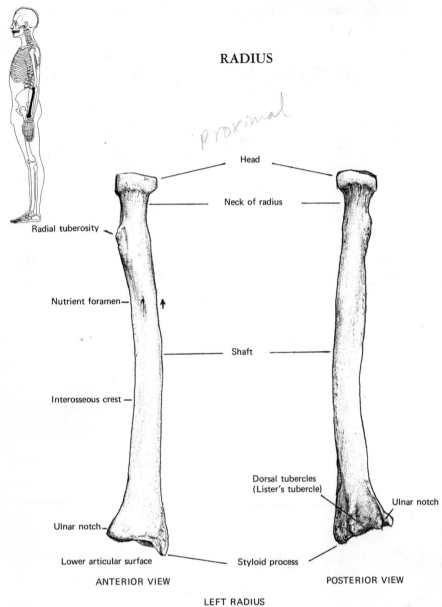

proximal

Head

Neck of radius

Radial tuberosity

Nutrient foramen

Shaft

Interosseous crest

Dorsal tubercles
(Lister's tubercle)

Ulnar notch

Ulnar notch

Lower articular surface

Styloid process

ANTERIOR VIEW

POSTERIOR VIEW

LEFT RADIUS

Fig. 70.

RADIUS: Paired—Long Bone (Figs. 70, 71)

Subadult bone:

The radius is ossified from a single center near the middle of the shaft which appears about the eighth week of intra-uterine life (Fig. 72). The distal epiphysis appears about age 1-1½ and unites in males about 17-18 and in females about 16-17.

120

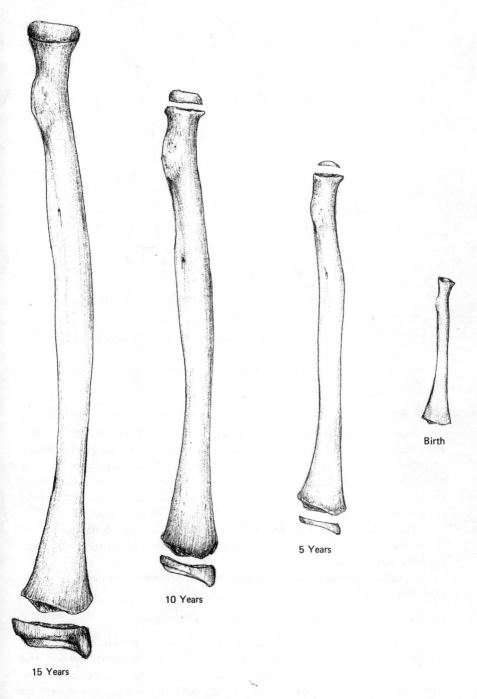

Birth

5 Years

10 Years

15 Years

Fig. 71.

121

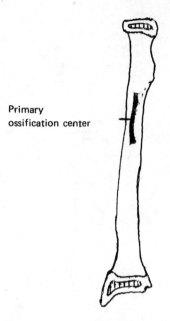

Primary
ossification center

Fig. 72.   OSSIFICATION OF RADIUS

The proximal epiphysis (head) appears about age 5-6 and unites about 16-18. The epiphysis for the head unites before the distal end.

Epiphyses at the elbow fuse about the 18th year, shoulder and wrist about the 20th year.

Adult bone:

The radius is the lateral and shorter of the two bones of the forearm. Proximally (head of the radius), it articulates with the humerus (at the elbow), and medially with the ulna. The distal end of the radius articulates with the navicular and lunate bones of the carpus (wrist), and medially with the ulna at the ulnar notch. The proximal extremity is smaller than the distal, and the head is a circular disk forming the expanded articular end of the bone.

The distal extremity is concave for articulation with the lunate bone, and the styloid process articulates with the navicular bone. The medial surface contains the ulnar notch.

Note that the nutrient foramen is inclined proximally (toward the head) (Fig. 66a).

BONES OF SIMILAR SHAPE WHERE CONFUSION MAY ARISE:

Ulna and *fibula* because of the comparable size of the shafts. The *radius* shaft is triangular with a prominent interosseous crest. The surface opposite this crest (lateral) is thick and rounded, and the triangular edges are not prominent. The *ulna* also has a triangular shaft with an interosseous crest, but the surface opposite the interosseous crest (medial) has sharper, more distinct edges. The *fibula* shaft tends to be irregular but more closely resembles the shaft of the ulna than it does the radius. The nutrient foramen in the ulna is larger and more prominent than in the fibula.

Side identification:

1. When the bone is held in approximate anatomical position with the head toward you, the nutrient foramen anterior and the distal end away from you, the styloid process is always on the lateral side and on the side the bone is from.

2. The interosseous crest and the radial tuberosity are always medial and on the opposite side the bone comes from. The ulnar notch (distal end) is also medial; thus, if it occurs on the right side of the bone you are holding, it is a left bone.

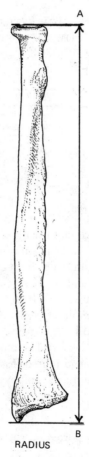

Fig. 73.   RADIUS

Measurements of the radius (Fig. 73):

Maximum length: (osteometric board)

Maximum length from head to tip of styloid process. Taken in same way as that of the humerus (A-B).

Humero-radial index: Useful for comparison of the humerus and radius.

$$\text{Humero-radial index} = \frac{\text{Maximum length of radius x 100}}{\text{Maximum length of humerus}}$$

Age estimation:

Johnston (1962:249-254) has published on the growth of long bones from the infants and children of the Indian Knoll, Kentucky skeletal population. The relation of age to length of the subadult bone is as follows where bone length is in millimeters.

TABLE 22. AGE ESTIMATION
FROM RADIUS
AFTER JOHNSTON'S TABLE 2
(1962:251)

Estimated age in years	Radius		
	n	Mean	σ
Fetal	5	47.20	5.42
NB–0.5	60	55.05	4.24
0.5–1.5	24	73.96	8.36
1.5–2.5	6	91.33	4.42
2.5–3.5	7	97.86	6.47
3.5–4.5	8	108.50	2.28
4.5–5.5	5	120.00	2.76

The epiphysis of the proximal end usually unites to the shaft about age 15-18. This union occurs about 2-4 years prior to union of the distal end which unites in females from 16-17 and in males from 17-19 (Greulich and Pyle 1959).

The epiphyses of the radius and ulna unite simultaneously and the data on age of union for the radius is similar to that for the ulna. McKern and Stewart (1957:47) report that in their sample of young American Korean war dead the proximal epiphyses had all united by age 19. Early stages of union for the distal epiphyses occurred from 17-20 in their male sample. By 23 years the distal epiphyses of both the radius and ulna had united. Last stages of union in the radius occur on the antero-lateral portion of the epiphyseal line.

Sex estimation:

A number of authors have reported on the difference in the proportions of upper to lower arm lengths as expressed in the Humero-radial index. However, few of these have been accurate enough to determine the sex and race of a single bone.

Ossification plays a major role in the sexing of the radius. The female tends to have ossification centers appearing earlier and complete union with the shaft terminating sooner than in the male. Studies by Garn, Rohmann, and Blumenthal found that in cases where roentgenograms are involved, the relationship between the radii capitulum of the radius and the medial epicondyle of the humerus can be used to determine sex in young specimens. There is a major sexual dimorphism with over 70% discriminatory efficiency in their particular ossification order. They found that the ossification order is capitulum radii-medial epicondyle in at least 75% of the males; 70% of the females had the opposite order. (Garn, Rohmann, and Blumenthal 1966:106).

Stature calculations:

The estimation of stature from long bones has been attempted by a number of authors. However, with the humerus, radius and ulna Trotter and Gleser say "it can be stated as a general rule that in no case should lengths of upper limb bones be used in the estimation of stature unless no lower limb bone is available" (1958:120).

Only those formulas given by Trotter and Gleser (1952, 1958) will be given. The radius usually gave the second from largest standard error of estimate (ulna the largest) of any of the long bones tested by Trotter and Gleser.

Stature formula for radius - male

White	Negro
3.79 Radius + 79.42 ± 4.66	3.32 Radius + 85.43 ± 4.57

Mongoloid	Mexican
3.54 Radius + 82.00 ± 4.60	3.55 Radius + 80.71 ± 4.04

When stature is estimated for an individual over 30 years of age the estimate should be reduced by the amount of 0.06 (age in years -30) cm.

Length of radius must be in centimeters

| Example | - for radius of white male that is 23.0 cm., (230mm.).

3.79 (23.0) + 79.42 ± 4.66
97.17 + 79.42 ± 4.66

176.59 cm. = mean
Range 176.59 - 4.66 = 171.93 cm. Low
176.59 + 4.66 = 181.25 cm. High

ULNA

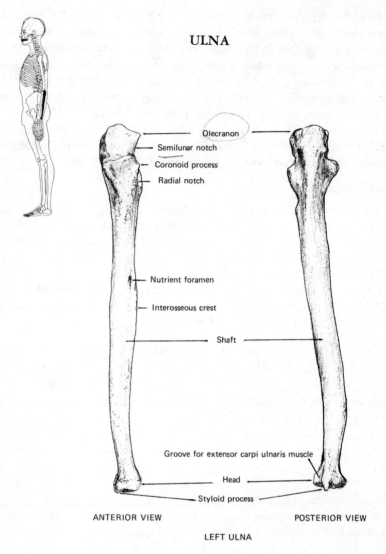

Olecranon

Semilunar notch

Coronoid process

Radial notch

Nutrient foramen

Interosseous crest

Shaft

Groove for extensor carpi ulnaris muscle

Head

Styloid process

ANTERIOR VIEW   POSTERIOR VIEW

LEFT ULNA

Fig. 74.

ULNA: Paired—Long Bone (Figs. 74, 75)

Subadult bone:

The ulna is ossified from a single center near the middle of the shaft which appears about the eighth week of intra-uterine life (Fig. 76). The distal epiphysis appears (ossifies) at age 6-7, or some 5-6 years later than the distal epiphysis of the radius. This epiphysis appears earlier than that at the proximal end but unites later—at about age 17-20.

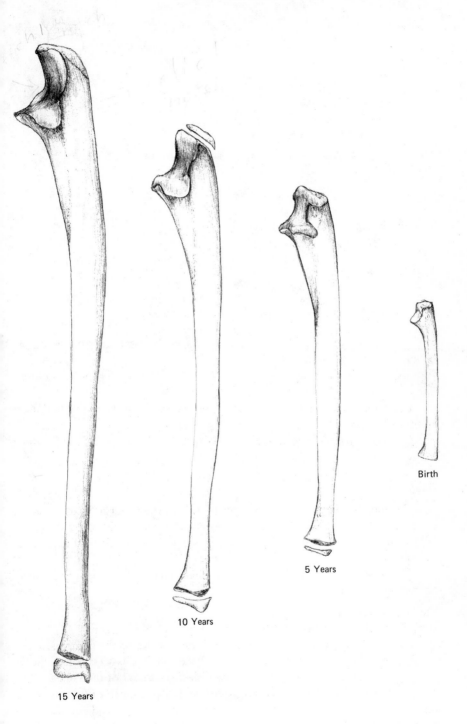

15 Years

10 Years

5 Years

Birth

Fig. 75.

127

Fig. 76.   OSSIFICATION OF ULNA

The epiphysis at the proximal end appears at about age 11 (7-14) and unites by the 19th year.

Epiphyses at the elbow fuse about the 18th year; shoulder and wrist about the 20th year.

Adult bone:

The ulna is a long bone on the medial side of the forearm. It articulates at the proximal end (elbow) with the humerus and laterally with the radius. Distally, it articulates with the radius, but connects only indirectly with the carpus (wrist).

The proximal extremity is of irregular shape and is the thickest and strongest part of the bone.

Proximal extremity:

1. The olecranon fits into the olecranon fossa of the humerus.
2. The semilunar notch articulates with the trochlea of the humerus.
3. The radial notch on the lateral side of the coronoid process is for articulation with the circumference of the disk-shaped head of the radius.

The shaft or body of the ulna is three-sided throughout much of its length, but tapers near the distal extremity, where it becomes smooth and rounded. The nutrient foramen is inclined toward the proximal end (as in the radius).

(See diagram on page 113 for direction).

Distal extremity:

1. This is small in size and consists of the anatomical "head" of the ulna. Note that this is at the opposite end of the bone from the head of the radius.
2. The styloid process projects from the medial and posterior part of the bone.

BONES OF SIMILAR SHAPE WHERE CONFUSION MAY ARISE:

Radius and *fibula* because of the comparable size of the shafts. The *radius* shaft is triangular with a prominent interosseous crest. The surface opposite this crest (lateral) is thick and rounded and the triangular edges are not prominent. The ulna also has a triangular shaft with an interosseous crest, but the surface opposite the interosseous crest (medial) has sharper, more distinct edges. The *fibula* shaft tends to be irregular, but more closely resembles the shaft of the ulna than it does the radius. The nutrient foramen in the ulna is larger and more prominent than in the fibula.

Side identification:

1. Holding the bone in approximate anatomical position with the proximal end toward you and the semilunar notch up, the radial notch (proximal end), the interosseous crest and the nutrient foramen will all be on the side the bone is from (Fig. 77a).
2. If you have just the distal end, hold the bone so that the styloid process is on top and the groove for the extensor carpi ulnaris will be on the side of the styloid process that the bone is from (Fig. 77b).
3. If you have only the shaft, locate the direction of the nutrient foramen (which is inclined proximally) and the interosseous crest will always be on the side the bone is from.

Fig. 77a.   LEFT ULNA   Fig. 77b.   LEFT ULNA

Measurements of the Ulna (Fig. 78):

Maximum length: (osteometric board)

Maximum length from the top of the olecranon to the tip of the styloid process. Taken in same way as that of the humerus (C-D).

Physiological length: (hinge caliper)

The two measuring points being the deepest point in the longitudinal ridge running across the floor of the semilunar notch (A), and the deepest point of the distal surface of the "head" (B), not taking the groove between it and the styloid process (A-B).

Least circumference of the shaft: (tape)

Located a little above the distal epiphysis, where the shaft, through the reduction of the muscular ridges and crests, becomes nearly cylindrical (E).

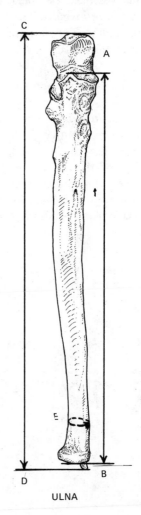

Fig. 78.   ULNA

Caliber index: This index expresses the relative delicacy or robustness of the bone as a whole—the greater the index the stouter the bone.

$$\text{Caliber index} = \frac{\text{Least circumference x 100}}{\text{Physiological length}}$$

Caliber indices of the ulna (after Wilder 1920:89)

	No.	
Gibbon	4	6.0
Orang	8	10.0
Gorilla	5	13.4
Chimpanzee	2	14.3
Australians	6	12.7
Melanesians	13	13.7
Negritoes	6	14.6
South Germans	25	16.8

Age estimation:

Johnston (1962:249-254) has published on the growth of long bones from the infants and children of the Indian Knoll, Kentucky skeletal population. The relation of age to length of the subadult bone is as follows where bone length is in millimeters.

TABLE 23. AGE ESTIMATION
FROM ULNA
AFTER JOHNSTON'S TABLE 2
(1962:251)

Estimated age in years	Ulna		
	n	Mean	σ
Fetal	5	54.80	4.12
NB–0.5	54	63.70	4.74
0.5–1.5	29	82.86	9.00
1.5–2.5	5	99.20	1.94
2.5–3.5	7	108.00	5.76
3.5–4.5	8	120.63	4.24
4.5–5.5	4	132.75	3.42

The epiphysis for the proximal end usually unites to the shaft about 15-18 in females, by the 19th year in males (McKern and Stewart 1957:47). This union occurs about 2-4 years prior to union of the distal end which unites in females from 15-16 and in males from 17-18 (Greulich and Pyle 1959).

Sex estimation:

Although many measurements and observations have been taken on the ulna, none have shown a high correlation with sex. In general, females are smaller and males are larger, but the overlap of the two populations is so great that little reliability is obtained from sex determination of the ulna alone.

Stature calculations:

The estimation of stature from long bones has been attempted by a number of authors. However, with the humerus, radius and ulna, Trotter and Gleser say "it can be stated as a general rule that in no case should lengths of upper limb bones be used in the estimation of stature unless no lower limb bone is available" (1958:120).

Only those formulas given by Trotter and Gleser (1952, 1958) will be given. The ulna usually gave the largest standard error of estimate of any of the long bones tested by Trotter and Gleser.

Stature formula for ulna – male

White	Negro
3.76 Ulna + 75.55 ± 4.72	3.20 Ulna + 82.77 ± 4.74

Mongoloid	Mexican
3.48 Ulna + 77.45 ± 4.66	3.56 Ulna + 74.56 ± 4.05

Length of ulna must be in centimeters.

Example for ulna of a white male that is 23.0 cm., (230 mm.).

3.76 (23.0) + 75.55 ± 4.72
86.48 + 75.55 ± 4.72

162.03 cm. = mean
Range 162.03 – 4.72 = 157.31 cm. Low
      162.03 + 4.72 = 166.75 cm. High

When stature is estimated for an individual over 30 years of age the estimate should be reduced by the amount of 0.06 (age in year -30) cm.

HAND

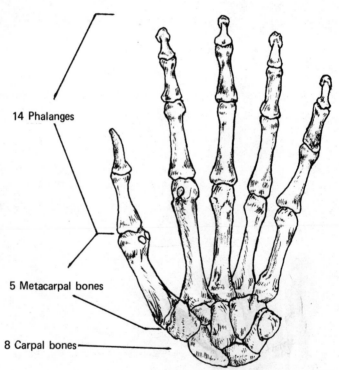

14 Phalanges

5 Metacarpal bones

8 Carpal bones

SKELETON OF LEFT HAND (DORSAL VIEW)

Fig. 79.

Carpal Bones: Paired—Irregular Bones (Figs. 79, 80)

The skeleton of the wrist consists of 8 carpal bones, arranged in two rows, with 4 bones in each row. From the thumb to the little finger, the proximal or first row consists of:

1. Navicular or scaphoid
2. Lunate
3. Triquetral
4. Pisiform

The distal or second row consists of:

5. Greater multangular or trapezium
6. Lesser multangular or trapezoid
7. Capitate
8. Hamate

All of the carpal bones, except the Pisiform, have 6 surfaces. The whole of the carpus is cartilaginous at birth and each bone is ossified from a single center.

Subadult bones:

At birth in the average individual the bones of the hand consist of the five metacarpals and 14 phalanges. No epiphyses are present and none of the carpal bones have begun to ossify.

At one year:

In males, the ossification of the capitate and hamate have begun. No epiphyses are present.

In females not only the capitate and hamate have begun to ossify but the distal epiphysis of the radius, the epiphyses of the second and third metacarpals, and the epiphyses of the proximal phalanges of the second, third and fourth fingers now contain small, recently formed ossification centers.

At five years:

In males, the epiphysis for the radius is about three-fourths as wide as the distal end of the shaft. There is no epiphysis for the ulna. All of the carpal bones have ossified except the Scaphoid (Navicular) and Trapezoid (Lesser Multangular). The epiphysis of the first metacarpal is more than half as wide as its metaphysis. The epiphyses of the metacarpals and proximal phalanges have appeared and enlarged. The epiphyses of all of the middle and distal phalanges have ossified.

In females all of the carpal bones have ossified. There is still no epiphysis on the distal end of the radius. All of the epiphyses of the metacarpals and phalanges are ossified and have expanded. The epiphyses of the middle phalanges of the second, third and fourth fingers are shaping to the contour of the trochlear surfaces of the proximal phalanges. The distal phalanges of the third, fourth and fifth fingers are now as wide as their shafts.

At 10 years:

In males the epiphyses for the radius, ulna, metacarpals and phalanges are ossified. The epiphyses of the phalanges are almost as wide as their shafts

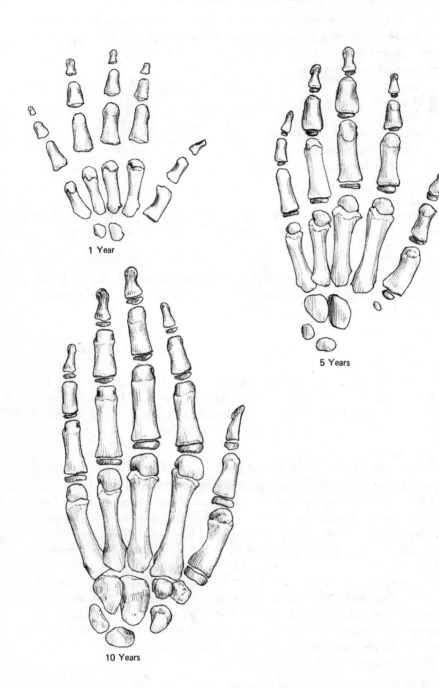

1 Year

5 Years

10 Years

Fig. 80.

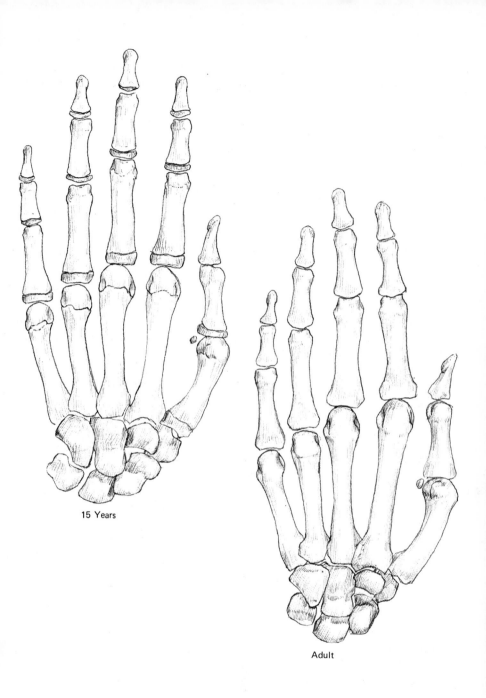

15 Years

Adult

Fig. 80. (cont.)

137

and those of the distal phalanges of the second to fifth fingers are all wider than their shafts.

In females the ossification of all epiphyses are more advanced than in the male. The styloid process of the ulna epiphysis is beginning to form. The epiphyses of the distal phalanges (especially the third finger) have begun to cap their shafts. In the following few years, most of the epiphyses of the phalanges, prior to their union with the shaft, will resemble pop bottle caps.

At 15 years:

In males the epiphysis of the radius has capped its shaft and that of the ulna is as wide as its shaft. All carpals have now attained their early adult shape. Fusion is underway in the epiphyses of all distal phalanges.

In females radial and ulna epiphyses have begun to fuse with their shafts progressing further in the ulna than in the radius. Fusion is completed or in the final stages of completion in all of the carpal epiphyses.

Fusion is completed first in the distal, next in the proximal, and last in the middle phalanges of the second, third, fourth and fifth fingers.

A detailed description of each bone through the aging process can be found in William W. Greulich and S. Idell Pyle's Radiographic Atlas of Skeletal Development of the Hand and Wrist, (1959) Stanford University Press, Stanford, California.

The following consists of general rules on how to tell right from left carpal bones.

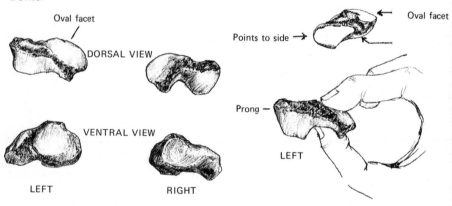

Fig. 81. Navicular

Navicular (Fig. 81)—The navicular is the largest of the proximal row, and is the first bone of the row on the thumb side. It articulates with the lateral facet on the distal end of the radius.

Side identification:

Put thumb in the hollow, forefinger on oval (radial) facet. Ridge side toward you and concave (cut out part or the notch where the two articular facets come together) toward you. The prong points to side it is from (articular facet for greater and lesser multangular).

138

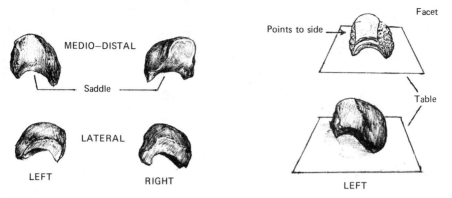

Fig. 82. Lunate

Lunate (Fig. 82)—This is the second bone in the proximal row.

Side identification:
   Place flat side down, saddle (concave articular surface for capitate) faces you and facet is on the side from which it comes.

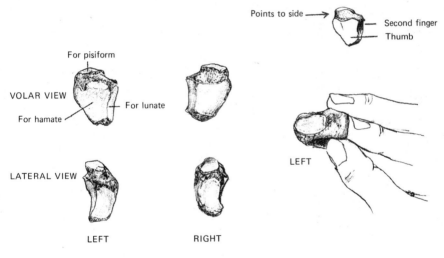

Fig. 83. Triquetral

Triquetral (Fig. 83)—This is the third bone from the thumb in the proximal row and is pyramidal in shape.

Side identification:
   There are 3 articular facets; 2 facets meet along a common edge (for lunate and hamate). With this edge held vertically and towards you (the thumb will be in the concave hamate facet), the third, and the only flat, facet (it joins the pisiform) is up, and points towards the side from which the bone comes. The right bone fits into the left hand better, and vice versa.

139

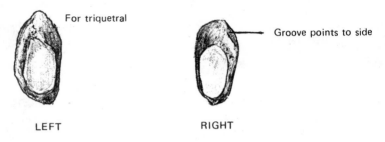

Fig. 84. Pisiform

Pisiform (Fig. 84)—This is the smallest of the carpal bones. It is the 4th and last bone from the thumb of the proximal row. In many of its characteristics, it is in complete contrast to the other bones. It has a single articular facet which is for the triquetral.

Side identification:

With facet on top of bone and tubercle away from you, the groove or depression immediately behind the edge of the articular surface is on side from which it comes. It should be noted that there is a considerable amount of variation in pisiforms, and some are difficult to side.

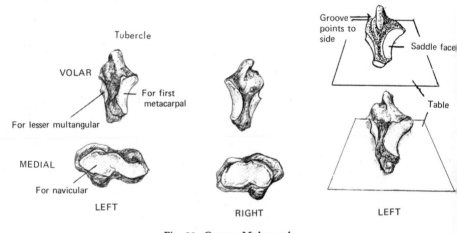

Fig. 85. Greater Multangular

Greater Multangular (Fig. 85)—This is the first bone on the distal carpal row, next to the thumb. It sits between the navicular and the first metacarpal.

Side identification:

Place on a flat surface with the 2 large articular (saddle) facets on either side. The tubercle will be on top, away from you, and will lean or form a lip over the groove on the side from which it comes.

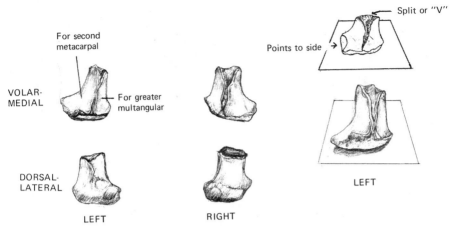

Fig. 86. Lesser Multangular

Lesser Multangular (Fig. 86)—This is the second and smallest of the carpal bones in the distal row.

Side identification:

The split (unarticular groove between articular facets for second metacarpal and greater multangular) goes toward you and the toe points toward the side from which the bone comes. Articular surfaces are on the opposite side.

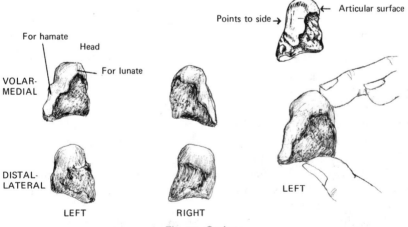

Fig. 87. Capitate

Capitate (Fig. 87)—Situated in the center of the wrist, this is the largest bone of the carpus. The rounded end is called the head.

Side identification:

With the head up (proximal rounded portion) and rough flat side toward you, the articular surface (for the hamate) narrows down toward the side from which it comes.

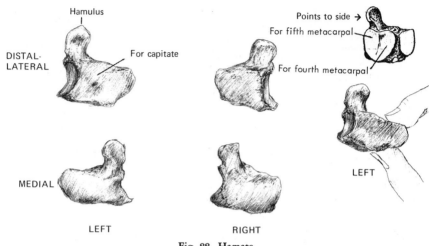

Fig. 88. Hamate

Hamate (Fig. 88)—This is a wedge-shaped bone bearing a hooklike process (the hamulus). It is the 4th and last of the carpal bones in the distal row.

Side identification:

Place the flat (non-articular), rough uneven surface down and hook (hamulus) away from you. With the large articular surface toward you, the hook is on the side from which it comes. The tip of the hamulus leans toward you.

Metacarpals: Paired—Short Bones (Fig. 89)

The metacarpals consist of five cylindrical bones in each hand. They articulate with the carpus proximally and with the first or proximal row of phalanges distally. They have been well described as long bones in miniature. As they extend from the carpus, they slightly diverge from each other. They are numbered from the lateral (or thumb) to the medial side.

As with long bones, the metacarpals present a shaft and two extremities. The base or carpal extremities articulate with the carpals proximally, and except for the first metacarpal (thumb), they articulate on the side with the adjacent metacarpal bones. This is a good method of distinguishing the first from the second through the fifth.

Note that the shafts are curved so as to be slightly concave on the volar (palm) surface of the hand, and thus slightly convex on the dorsal side (back). They are triangular in cross section.

The head or distal extremity articulates with the proximal (first) row of phalanges. Note that the heads all present a large rounder articular surface, extending further on the volar than on the dorsal aspect. This enables one to clench one's fist. Although the nutrient foramina are difficult to find in many metacarpals, it should be noted that they are inclined toward the distal end of the first, but toward the proximal end of the second through the fifth.

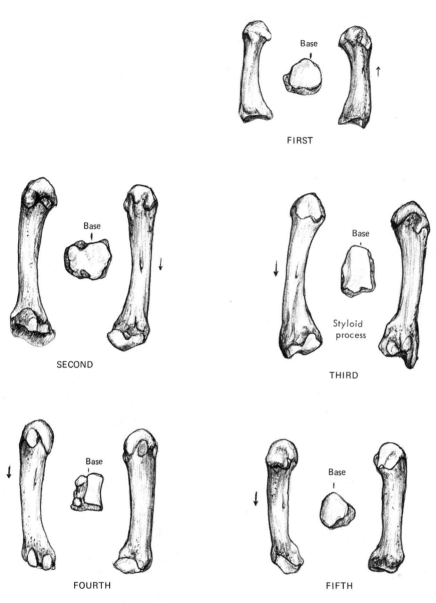

FIRST

Base

SECOND

THIRD

Base

Styloid
process

FOURTH

FIFTH

Base

ALL ARE LEFT BONES WITH RADIAL, BASE, AND ULNAR VIEWS

Fig. 89. Metacarpals

First (thumb)—shortest and thickest of the metacarpals; it has a saddle-shaped carpal articular surface.

Second—longest metacarpal.

Third, fourth and fifth successively decrease in length.

Side determination:

The description for the determination of the metacarpals is based upon all bones being held with the dorsal (back) surface toward you:

First—both the proximal and distal ends project further on the side the bone is from (Fig. 90a). Has a saddle-shaped proximal articular surface.

Fig. 90a.

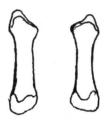

Second—the proximal end projects further on the side the bone is from (Fig. 90b). Head is inclined toward the opposite side.

Fig. 90b.

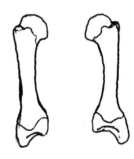

Third—the proximal end (styloid process) projects further *OPPOSITE THE SIDE* the bone comes from (Fig. 90c).

Fig. 90c.

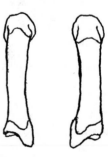

Fourth—the entire proximal end is inclined toward the side the bone is from (Fig. 90d).

Fig. 90d.

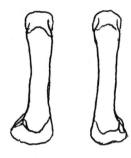

Fifth—articular facet on only the lateral surface of shaft presents an oblique line which is inclined down the bone toward the proximal end and off to the side from which the bone comes (Fig. 90e). Also when bone is held with dorsal surface toward you, the lateral articular facet is opposite the side the bone comes from.

Fig. 90e.

Differences between metacarpals (hand) and metatarsals (foot) (Fig. 90f).
There are the same number (5) in both the hand and the foot. —
1. The metatarsals are longer and the shafts (except for the first) are a little thinner than the metacarpals.
2. The articular surfaces on the heads of the metatarsals are restricted laterally and are well marked anteriorly-posteriorly. This gives the appearance of having a groove causing a double expansion of the head.
3. The grooves between the articular facets on the proximal end of the metatarsals are much more pronounced. They are deeper and more rugged than on the metacarpals.

Fig. 90f.

METACARPAL   METATARSAL

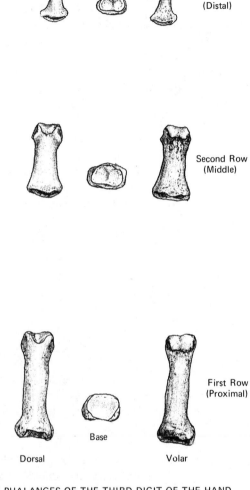

Third Row
(Distal)

Second Row
(Middle)

First Row
(Proximal)

Base

Dorsal

Volar

Fig. 91.　THE PHALANGES OF THE THIRD DIGIT OF THE HAND
THREE VIEWS: DORSAL, BASE, AND VOLAR

Phalanges (Fingers): Paired—Short Bones (Fig. 91)

There are fourteen phalanges in each hand, two for the thumb and three for each of the other digits. They are divided into three rows:

First or proximal row—5; these are the largest

Second or middle—4 (no middle phalanx in the thumb)

Third or distal or terminal—5; these are the smallest

In all of the phalanges, the nutrient foramen is directed toward the distal extremity. They are not easily seen. The phalanges present a shaft and two extremities.

First or proximal row: The proximal (metacarpal) extremity presents a single concave, oval articular surface which receives the convex head of the metacarpal bone. The distal extremity is grooved in the center and elevated on each side into two little condyles. When the hand is closed, the distal ends of the proximal and middle phalanges are uncovered and can be felt. The two small condyles on the distal extremity of the proximal and middle phalanges resemble the lower end of the femur. To correspond with these condyles, the bases of the terminal and middle phalanges have two small depressions, and resemble the proximal end of the tibia.

Second or middle row: The base or proximal extremity presents two shallow depressions separated by a medial ridge. The distal end articulates with the base of the third phalanx and is grooved in the center and elevated on each side into two small condyles.

Third or distal row: Small in size, they are easily recognized because the distal end is tapered (they are neither weight bearing or force-transmitting). The dorsal side is smooth over which the fingernail fits, and the volar is rough because of the attachment of the fiber bands that attach the finger pad to it.

The proximal end is similar in shape with that of the second phalanx in that it presents two shallow depressions separated by a medial ridge. One could never confuse the phalanges of the middle and distal row. Both ends of the middle phalanges have articular surfaces whereas only the proximal end of the phalanges in the distal row have an articular surface.

Differences between the phalanges of the hand (fingers) and foot (toes):
There are the same number of phalanges (14) in both the hand and foot.
1. With the exception of the two phalanges of the big toe, which are larger than the two phalanges of the thumb, the remaining 12 phalanges of the toes are smaller and more rudimentary than the corresponding bones in the fingers.
2. The phalanges of the fingers are flat on the volar (palm) surface and rounded on the dorsal surface (Fig. 92).
3. The shaft is narrow and compressed in the phalanges of the toes and the bones are usually not as long as in the fingers.

The nutrient foramina, which are difficult in many bones to locate, extend toward the distal end in both the fingers and toes.

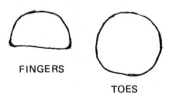

FINGERS

TOES

Fig. 92.

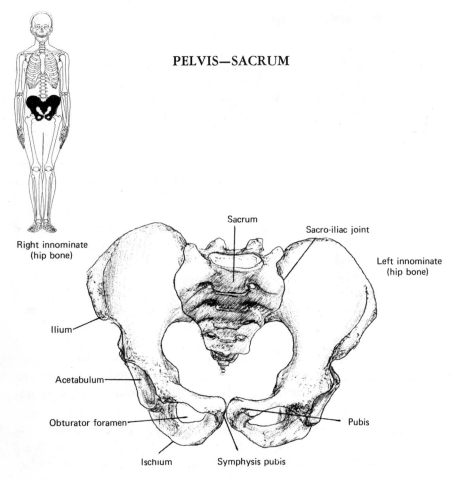

Right innominate
(hip bone)

Sacrum

Sacro-iliac joint

Left innominate
(hip bone)

Ilium

Acetabulum

Obturator foramen

Pubis

Ischium

Symphysis pubis

FEMALE PELVIS — VENTRAL VIEW

Fig. 93.

HIP BONE (INNOMINATE): Paired—Irregular Bones (Figs. 93-96)

Subadult bone:

The hip bone consists of three distinctive portions that unite about the 12th year (Fig. 97). Prior to adolescence the hip bone consists of three separate bones, the ilium for which an ossification center appears about the 2nd-3rd month of intra-uterine life, the ischium ossifying about the 4th month and the pubis about the 5th month.

The rami of the pubis and ischium fuse in the 7-8 year. In the 12th year the cartilaginous strip at the acetabulum which has separated the three bones since birth begins to ossify. Complete fusion may last as late as 17 McKern and Stewart (1957:57).

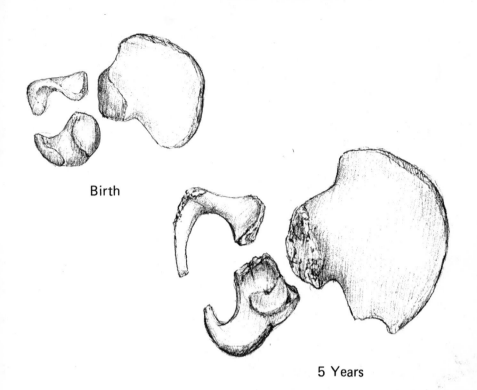

Birth

5 Years

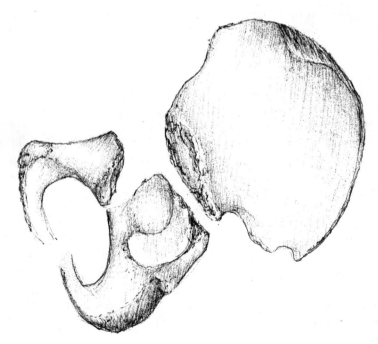

10 Years

Fig. 94.

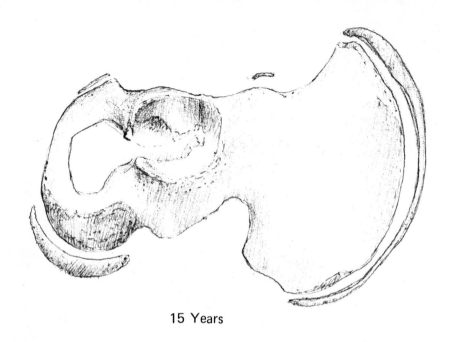

15 Years

Epiphyses, the iliac crest, the anterior inferior iliac spine, the pubis and the ischial tuberosity appear about puberty and unite from 16-23. Union is quite variable, as pointed out by McKern and Stewart (1957:57-71). They found the iliac crest to be completely united by age 23 in males, and that of the ischial tuberosity to be completely fused by 24.

Adult bone:

The hip bone is a large irregular bone which, with its mate of the opposite side and the sacrum, go together to make up the pelvis. It is sometimes called the innominate bone because it bears no resemblance to any common object. This is one of the most difficult bones of the body for students to learn. It is difficult for the beginner to orient this bone properly and there are a number of spines, foramina and notches to be learned. If the proper orientation of the bone is learned there should be little difficulty in side determination.

Each bone consists of three parts which, though separate in early life, are united into one bone in adulthood:

    1. Ilium—upper portion of hip bone.
    2. Ischium—lower portion of hip bone; supports the body in sitting position.
    3. Pubis—anterior portion of hip bone; articulates with the opposite hip bone at the anterior midline of the body at the pubic symphysis.

The three separate bones join into the formation of the acetabulum (joint for the femur), with the ilium and ischium contributing approximately ⅖ each and the pubis ⅕ of the acetabulum (see lateral view of hip bone). Note that the acetabulum has both articular and nonarticular portions.

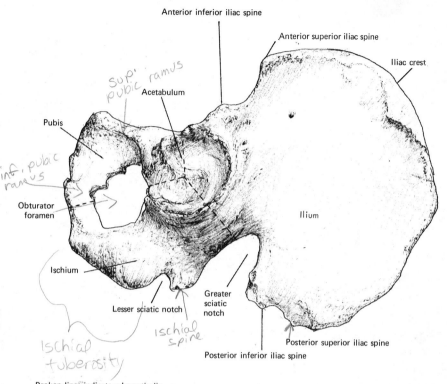

Anterior inferior iliac spine

Anterior superior iliac spine

Iliac crest

sup' pubic ramus

Acetabulum

Pubis

inf, pubic ramus

Obturator foramen

Ilium

Ischium

Greater sciatic notch

Lesser sciatic notch

Ischial spine

Ischial tuberosity

Posterior superior iliac spine

Posterior inferior iliac spine

Broken lines indicate schematically
the lines of union of the three parts of the innominate.

LEFT HIP BONE (LATERAL VIEW)

Fig. 95.

BONES OF SIMILAR SHAPE WHERE CONFUSION MAY ARISE:

Scapula and *flat bones of the skull.* The scapula is most often thinner and presents sharper borders. Students sometimes mistake a fragmentary ilium for the flat bones of the skull. Remember that the cranial bones have suture lines; these are not found in the hip bone.

Side identification:

1. Holding the bone in approximate anatomical order, the ilium will be posterior, the pubis anterior and the greater sciatic notch down (inferior). The acetabulum will be on the outside (lateral) (for the articulation of the femur), and thus is on the same side the bone is from.

2. The articular surface of the ilium (sacro-iliac joint) is inside (medial) and is always posterior to the greater sciatic notch. The pre-auricular sulcus

151

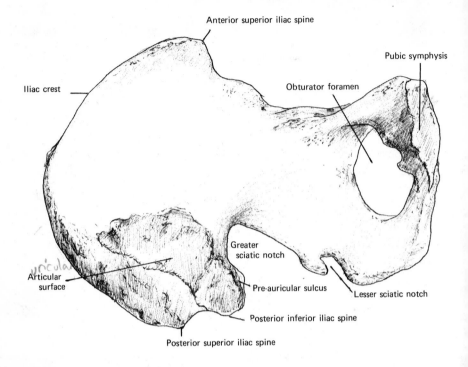

Fig. 96.   LEFT HIP BONE (MEDIAL VIEW)

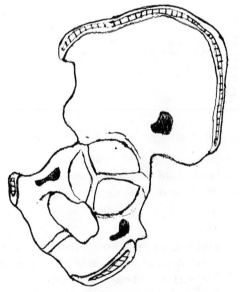

Fig. 97.   OSSIFICATION OF THE HIP BONE

152

(well defined in females) is just anterior to the articular surface of the ilium and is between the articular surface and the greater sciatic notch.

3. The pubis is smooth on the medial (back) surface and is rougher on the ventral (front) surface.

Measurements of the innominate (Fig. 98):

Maximum height: (osteometric board).

Place the ischial tuberosity (or ischium) against the fixed vertical of the board and the movable upright to the iliac crest. Raise the bone slightly and move it up and down as well as from side to side until the maximum length is obtained (A-B).

Maximum breadth: (osteometric board or sliding calipers).

The distance between the anterior-superior iliac spine to the posterior-superior iliac spine (C-D).

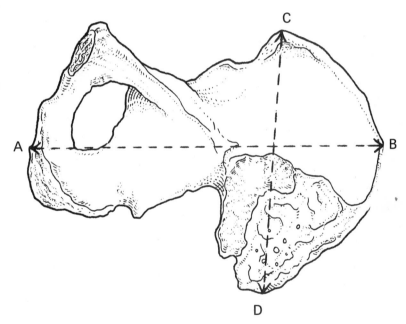

Fig. 98.

Ischium-Pubis index:

An index was used by Washburn (1948) in order to easily and effectively measure the difference in proportion between male and female pelves. The measurement of the sub-pubic angle is often made for this same reason; however, both innominates are necessary for this measurement. The ischium-pubis index can be calculated from one innominate alone.

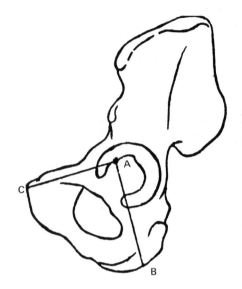

Fig. 99. Outline of left innominate. Point A is located in the acetabulum, the cup-shaped cavity receiving the head of the femur. This point represents the meeting place of the three elements from which the innominate is formed. The pubis is one of these elements and its length is measured by the line $A-C$. Another of these elements is the ischium, and its length is measured by the line $A-B$. The sex of the specimen is indicated by the relationship of these two lines, according to a formula given in the accompanying text. (After Schultz, 1930.)

The length of the ischium and pubis is measured from the point at which they meet in the acetabulum (Washburn 1948:200). This meeting point is characterized by an irregularity, change in thickness of the bone, or a notch. As Stewart (1954:418) has pointed out, the location of point A is difficult. Point A represents the meeting place of the three elements that unite to form the innominate (Fig. 99).

$$\text{Ischium-pubis index:} \quad \frac{\text{Pubis length x 100}}{\text{Ischium length}}$$

The ischium-pubis index aids in sex determination according to its position as follows:

Whites (N= 200)	Negroes (N= 100)
Below 90 = Male	Below 84 = Male
90-95 = Sex?	84-88 = sex?
95+ = Female	88+ = Female

This index averages 15% higher in females than males (Washburn 1948: 206). Montagu (1960: 629) offers the following information concerning statistics for this index according to sex:

	Mean	Range
White Male	83.6 ± 4.0	73 - 94
White Female	99.5 ± 5.1	91 - 115
Negro Male	79.9 ± 4.0	71 - 88
Negro Female	95.0 ± 4.6	84 - 106

Age estimation:

The innominate is probably the most important bone in age determination. There are many changes from the subadult bone (see sub-adult section) to the changes in the pubic symphysis of adults.

Pubic symphysis.

Following the union of the epiphyses, one of the best areas to determine age of an adult is from the pubic symphysis, the adjoining areas where the two hip bones come together in front. Todd (1920) observed that the symphyseal face of the pubic bone undergoes a regular metamorphosis from puberty onward. He established ten phases of pubic symphysis age, ranging from 18 to 50+ years (Plate 5). These phases as defined by Todd (1920) are as follows:

I. *First Post-adolescent:* 18-19 years. Symphysial surface rugged, traversed by horizontal ridges separated by well marked grooves; no ossific nodules fusing with the surface; no definite delimiting margin; no definition of extremities (p. 301).

II. *Second Post-adolescent:* 20-21 years. Symphysial surface still rugged, traversed by horizontal ridges, the grooves between which are, however, becoming filled near the dorsal limit with a new formation of finely textured bone. This formation begins to obscure the hinder extremities of the horizontal ridges. Ossific nodules fusing with the upper symphysial face may occur; dorsal limiting margin begins to develop, no delimitation of extremities; foreshadowing of ventral bevel (pp. 302-303).

III. *Third Post-Adolescent:* 22-24 years. Symphysial face shows progressive commencing formation of the dorsal plateau; presence of fusing ossific nodules; dorsal margin gradually becoming more defined; beveling as a result of ventral rarefaction becoming rapidly more pronounced; no delimitation of extremities (p. 304).

IV. *25-26 Years:* Great increase of ventral beveled area: corresponding delimitation of lower extremity (p. 305).

V. *27-30 Years.* Little or no change in symphysial face and dorsal plateau except that sporadic and premature attempts at the formation of a ventral rampart occur; lower extremity, like the dorsal margin, is increasing in clearness of definition; commencing formation of upper extremity with or without the intervention of a bony (ossific) nodule (p. 306).

VI. *30-35 Years.* More difficult to appraise correctly; essential feature is completion of oval outline of symphysial face. More individual variation than at younger ages; and terminal phases affect relatively minor details. Also, tendency for terminal phase to be cut short. Increasing definition of extremities; development and practical completion of ventral rampart; retention of granular appearance of symphysial face and ventral aspect of pubis; absence of lipping of symphysial margin (p. 308).

VII. *35-39 Years.* Paramount feature: face and ventral aspect change from granular texture to fine-grained or dense bone. Changes in symphyseal face

155

and ventral aspect of pubis consequent upon diminishing activity; commencing bony outgrowth into attachments of tendons and ligaments, espethe gracilis tendon and sacro-tuberous ligament (p. 310).

VIII. *39-44 Years.* Symphysial face generally smooth and inactive; ventral surface of pubis also inactive; oval outline complete or approximately complete; extremities clearly defined; no distinct "rim" to symphysial face; no marked lipping of either dorsal or ventral margin (p. 311).

IX. *45-50 Years.* Characterized by well-marked "rim." Symphysial face presents a more or less marked rim; dorsal margin uniformly lipped; ventral margin irregularly lipped (p. 312).

X. *50 Years +.* Rarefaction of face and irregular ossification. Symphysial face eroded and showing erratic ossification; ventral border more or less broken down; disfigurement increases with age (p. 313).

McKern and Stewart (1957) have a much more objective system for studying the symphyseal surface but it is quite complicated and difficult for the unskilled to use. Their system should be consulted when age determination is required.

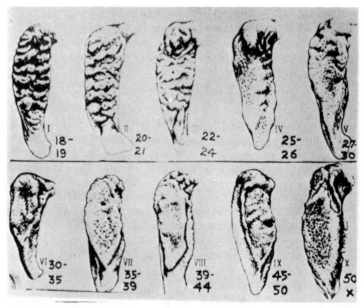

Plate 5. Modal standards of Todd's 10 typical phases. From Krogman, 1962, The Human Skeleton in Forensic Medicine. Courtesy of Charles C. Thomas, Publisher, Springfield, Ill.

Sex determination:

As in the living, the best area to tell the sex of a skeleton is from the pelvis. The highest accuracy has been achieved using this bone. (Genoves 1959; Krogman 1962; Phenice 1967; Washburn 1948.)

By the time an individual has reached the level of academic achievement that he is able to read this manual, he has already observed that the female has a broader pelvis (hips) than the male. This can be observed especially in spring and summer when coats have been replaced by more tightly fitting apparel. This greater width in the female pelvis is due to changes in three basic areas of the pelvis.

Male   Female

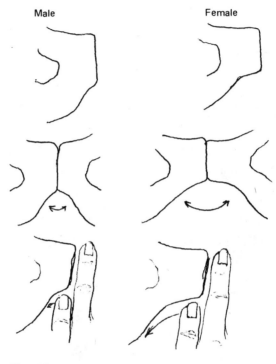

Fig. 100.

I. Pubic bone (Fig. 100)

Females have a longer pubic portion of the hip bone. Therefore, the sub-pubic angle is greater in females.

As a "rule of thumb", when the index finger is held perpendicular to the pubic symphysis the thumb can be moved only slightly, if at all, with a male innominate, but has ample room for movement on a female innominate.

II. Attachment of the arcuate ligament

There are three characteristics of the female pubis and ischiopubic ramus which serve to distinguish the sexes over 95% of cases (Phenice 1967:297-301).

1. The Ventral Arc: This is a slightly elevated ridge of bone which takes a course across the ventral surface of the female pubis (Fig. 101a).

Elevated Ridge
(Ventral Arc)

Fig. 101a.

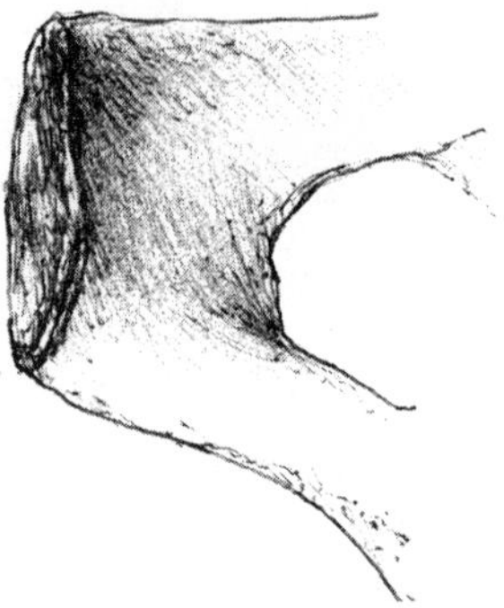

Subpubic Concavity

Fig. 101b.

2. The Subpubic Concavity: A lateral curvature a short distance inferior to the symphysis in the *female*. This is best observed from the dorsal surface of the bone (Fig. 101b).

3. The Medial Aspect of the Ischiopubic Ramus: In the *female* the medial aspect of the ischiopubic ramus presents a ridge or a narrow surface immediately below the symphyseal surface (Fig. 101c).

Narrow Medial Aspect

Fig. 101c.

In the male, there is no ventral arc. A ridge may appear on the ventral surface. If it does it will take one of the two forms shown. A ridge on the male pubis rarely, if ever appears as the ventral arc of the female. This is particularly true if the pubis is oriented for proper observation, with the ventral surface directly facing the observer and the symphyseal surface in a direct anterior-posterior plane (Fig. 101d).

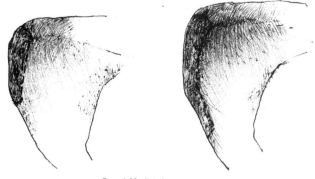

Fig. 101d.   Broad Medial Aspect

In the male there is rarely a subpubic concavity.

In the male the medial aspect of the ischiopubic ramus is a broad surface (Fig. 101e).

Fig. 101e.   Left Ventral View

When one or two of these three criteria are ambiguous, there will almost always be one that is definitely male or female. In such cases sex estimation should be based on the most distinctive criterion in the particular specimen (Fig. 102). The presence or absence of the ventral arc probably carries the most weight of the three criteria.

III. Sciatic Notch

The sciatic notch is wide in females and narrow in males. Another "rule of thumb" is to place your thumb in the sciatic notch. If the notch is filled or there is only limited side to side movement possible, it is a male. If considerable side to side movement is possible it is a female (Fig. 103a).

IV. Sacro-iliac Joint

In many innominate bones the area of the sacro-iliac articulation is raised in females (Fig. 103b).

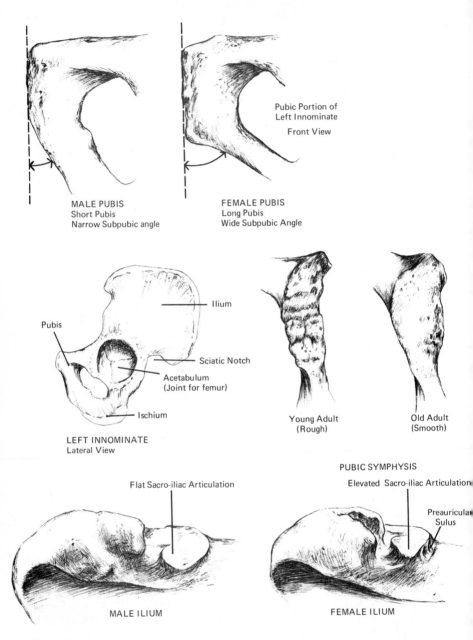

MALE PUBIS
Short Pubis
Narrow Subpubic angle

FEMALE PUBIS
Long Pubis
Wide Subpubic Angle

Pubic Portion of
Left Innominate

Front View

Pubis

Ilium

Sciatic Notch

Acetabulum
(Joint for femur)

Ischium

LEFT INNOMINATE
Lateral View

Young Adult
(Rough)

Old Adult
(Smooth)

PUBIC SYMPHYSIS

Flat Sacro-iliac Articulation

Elevated Sacro-iliac Articulation

Preauricular
Sulus

MALE ILIUM

FEMALE ILIUM

Fig. 102.

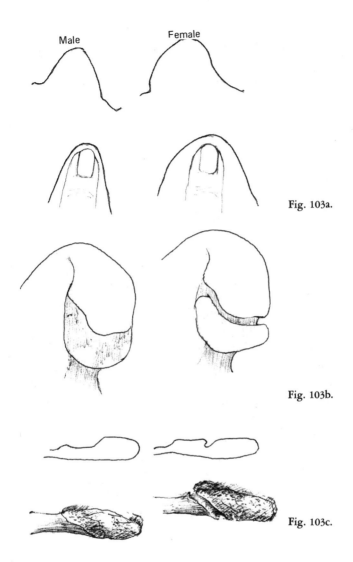

Fig. 103a.

Fig. 103b.

Fig. 103c.

V. Pre-auricular Sulcus

The pre-auricular sulcus is a depression between the sciatic notch and the sacro-iliac articulation. It is most often found in females (Fig. 103c).

VI. Other observational methods

There are additional morphological observations which help in estimating the sex, but they are generally felt to be of minor value (Fig. 104).

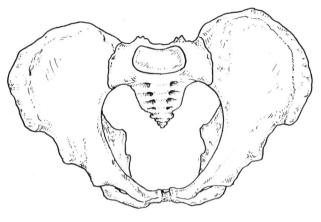

Male pelvis

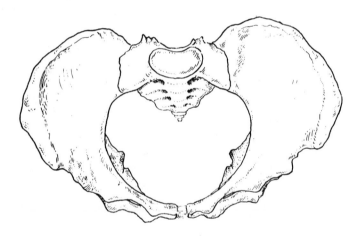

Female pelvis

Fig. 104.

Some are as follows:
1. In general, the male pelvis is more robust and muscle-marked.
2. The obturator foramen is larger and oval in males, whereas it is smaller and more triangular in females.
3. Since the female pelvis is adapted for childbirth, the pelvic basin is more spacious and less funnel-shaped (see pelvic drawings).
4. The acetabulum is larger in males to accommodate the larger femoral head.

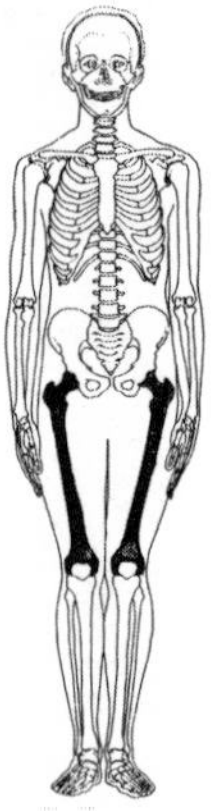

FEMUR

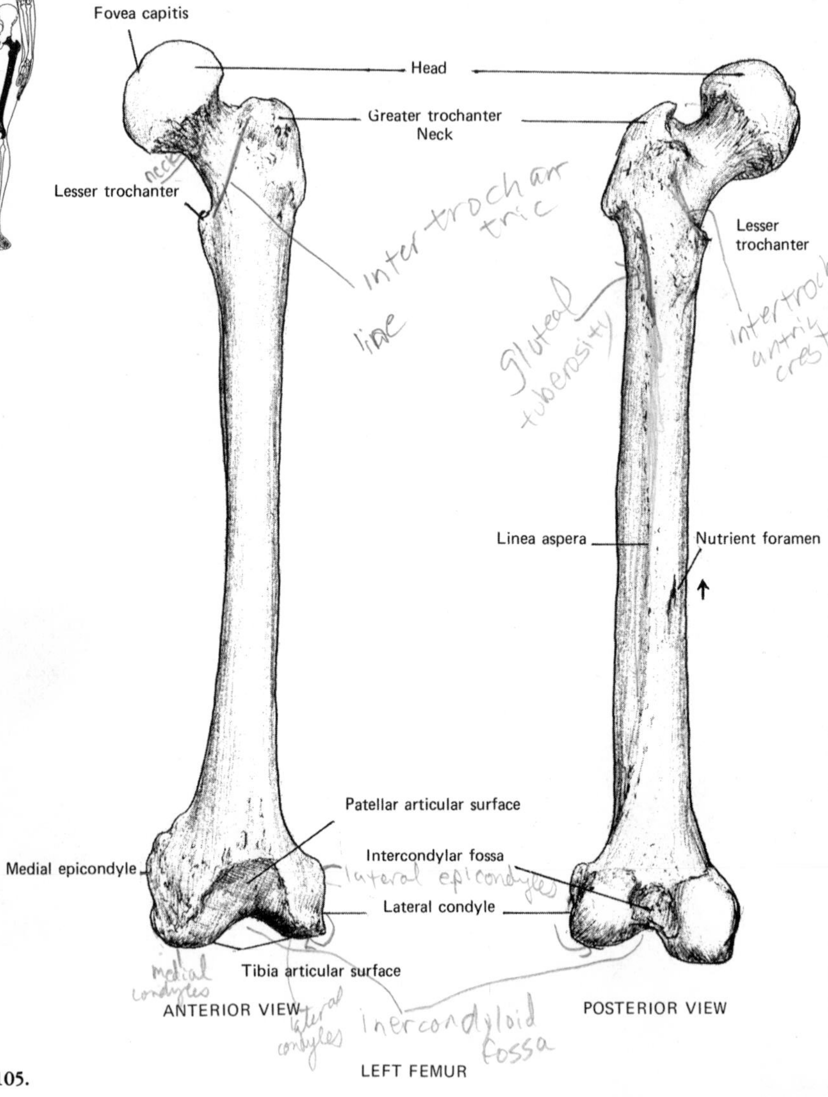
Fovea capitis
Head
Greater trochanter
Neck
Lesser trochanter
Lesser trochanter
Linea aspera
Nutrient foramen
Patellar articular surface
Medial epicondyle
Intercondylar fossa
Lateral condyle
Tibia articular surface
ANTERIOR VIEW
POSTERIOR VIEW
LEFT FEMUR

Fig. 105.

FEMUR: Paired—Long Bone (Figs. 105, 106)

Subadult bone:

The femur is ossified from a primary center for the shaft which appears about the 8th week of intra-uterine life and from four epiphyseal centers (Fig. 107).

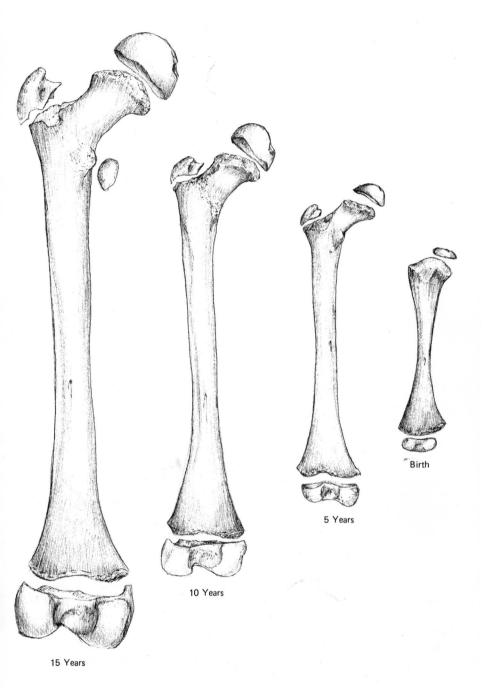

15 Years

10 Years

5 Years

Birth

Fig. 106.

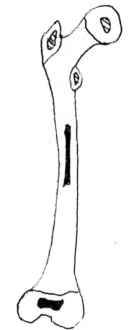

Fig. 107. OSSIFICATION OF THE FEMUR

The distal epiphysis begins to ossify before birth (this is usually the only secondary center to appear before birth). The nucleus for the head appears at about one year. Those for the greater trochanter appear about the 4th year and for the lesser trochanter about the 11th year.

The epiphyses at the proximal (head) end of the femur unite before that of the distal end. The epiphyses for the head, greater and lesser trochanter unite between 14 and 19. McKern and Stewart (1957:48) found 80% of their young male samples were united by 18. The distal epiphysis unites from 14 (females) to 18 (Pyle and Hoerr 1955). McKern and Stewart state that for their male sample "ossification is still in early stages as late as 20 years and does not become complete for all cases until the 22nd year" (1957:48).

Adult bone:

The femur is the largest and longest bone in the skeleton, and articulates with the hip bone superiorly and with the tibia inferiorly (at the knee). The heads of the two femora are widely separated by the pelvis but the femur in erect posture inclines from above down, slightly backwards and medially so that the distal ends are fairly close together at the knee.

The superior extremity, the head, is smooth for articulation, and forms two-thirds of a sphere to allow for great mobility at the hip joint.

The linea aspera, a longitudinally projecting ridge on the posterior surface is for attachment of muscles used for upright stature, and is found only in man.

It is possible for a student with little experience to confuse the shafts of the *humerus* and *tibia* with the femur, but a little practice with any of the three will soon solve this problem. The femur, as the largest bone of the body, has a larger shaft than the *humerus* and has a linea aspera. No comparable line is found on the humerus shaft. The femur shaft is smooth and rounded whereas the *tibia* shaft is triangular.

Side identification:

1. When holding the bone in approximate anatomical position, the head of the femur is superior and is opposite the side the bone is from (on the medial side) (Fig. 108a).
2. If you have only the shaft, locate the nutrient foramen and note that it is inclined proximally (toward the head) and is on the posterior surface of the shaft. Holding the shaft's posterior surface toward you and proximal end away from you, the linea aspera descends the bone and is inclined off to the side the bone is from (Fig. 108a).
3. If you have just the distal end, hold the anterior surface toward you (intercondylar fossa away from you). The patellar articular surface comes farther up the shaft (toward your hand) on the same side the bone is from (Fig. 108b).

Fig. 108a.   LEFT FEMUR

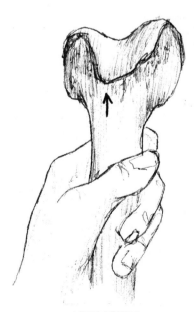

Fig. 108b.   LEFT FEMUR

4. Note that the head of the femur has a pit in the articular surface called the fovea capitis. Compare this with the head of the humerus which has no fovea. This is an excellent way of determining from a fragmentary head whether it is humerus or femur (Fig. 108c).

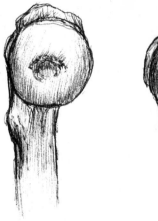 

Fig. 108c. FEMUR HUMERUS

Measurements of the femur (Fig. 109):

The femur is one of the most measured and reported on bones of the body. Ingalls (1924) lists 35 measurements on the femur alone. The following are basic measurements only.

Maximum length: (osteometric board).

Place the distal condyles against the fixed vertical of the board and the movable upright to the head. Raise bone slightly and move up and down as well as from side to side until maximum length is obtained (A-B).

Bicondylar (oblique or physiological length): (osteometric board).

Place both condyles against the fixed upright of the board and with the bone lying on the board apply movable end to the head (C-D).

Anterior-posterior diameter of the mid-shaft: (sliding caliper).

Locate mid-point of shaft on osteometric board and mark bone with pencil. Measure maximum anterior-posterior diameter (S-T).

Medio-lateral diameter of the mid-shaft: (sliding caliper).

Taken at right angles to the previous measurement. The linea aspera should be midway between the two branches of the caliper (M-N).

Maximum diameter of the head: (sliding caliper).

Measured on the periphery of the articular surface of the head. Rotate the bone until the maximum distance is obtained (E-F).

Circumference of the mid-shaft: (tape).

Taken at the middle of the shaft.

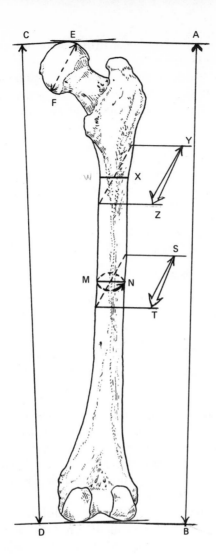

Fig. 109.

Subtrochanteric anterior-posterior diameter: (sliding caliper).

Taken on the shaft just below the lesser trochanter, with the gluteal tuberosity avoided; this measurement gives the minimum diameter of flattening (Y-Z).

Subtrochanteric medio-lateral diameter: (sliding caliper).

Taken at the same level as the previous measurement but perpendicular to it; this gives the maximum lateral diameter of flattening (W-X).

Platymeric index:

The proximal part of the shaft of the femur shows considerable difference in general shape between various populations.

Platymeric index: $\dfrac{\text{Subtrochanteric anterior-posterior diameter x 100}}{\text{Subtrochanteric medio-lateral diameter}}$

Range: Platymeric—X-84.9—broad or flat (from front to back)
     Eurymeric—85.0-99.9
     Stenomeric—100.0-X—(usually found only in pathological cases)
     Brothwell (1963) has a short but good discussion on the possible cause of differences in the platymeric index and gives the following degrees of known variability.

Fossil man:	Cro-Magnon man	73*
	Neanderthal man	77
Recent groups:	Turks	73
	American Indians	74
	Andamanese	78
	Eskimo	81
	Australians	82
	English (17th Cent.)	85

*From Brothwell (1963) Table 2, page 91.

Robusticity index: This expresses the relative size of the shaft.

Robusticity index:

$$\frac{\text{Anterior-posterior} + \text{medio-lateral diameter of mid-shaft} \times 100}{\text{Bicondylar (Physiological) length}}$$

Age estimation:

Fetal—The femur is one of the few bones of the skeleton upon which measurements of maximum length have been reported for the fetal skeleton.

Often, especially in archaeological material, the diaphysis or shaft is present but the epiphyses are not. When faced with the problem of estimating the age at death from a diaphysis only (do not include epiphyses), the following two figures from Stewart (1954) are helpful.

After birth, Johnston (1962:249-254) has published on the growth of long bones from the infants and children of the Indian Knoll, Kentucky skeletal population (Table 24). The relation of age to length of the subadult bone is as follows where the bone length is in millimeters.

TABLE 24. AGE ESTIMATION
FROM FEMUR
AFTER JOHNSTON'S TABLE 2
(1962: 251)

| Estimated age in years | Femur | | |
	n	Mean	σ
Fetal	7	61.86	6.33
NB–0.5	64	78.84	7.23
0.5–1.5	38	115.63	18.34
1.5–2.5	8	148.13	10.76
2.5–3.5	11	166.73	9.99
3.5–4.5	11	183.82	9.20
4.5–5.5	6	213.67	4.52

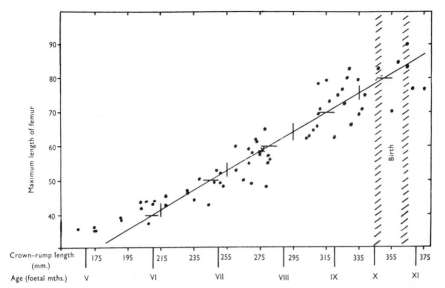

Fig. 110. Distribution of maximum dried femur lengths in 65 foetal skeletons, plotted against crown-rump length as obtained in the same specimens before maceration. The ages in foetal months corresponding to the crown-rump lengths (Scammon, 1937) are also charted. By locating on the regression line the length of any femur belonging to this size range, the approximate age may be found directly below. From Stewart, 1968, Fig. 53, p. 132.

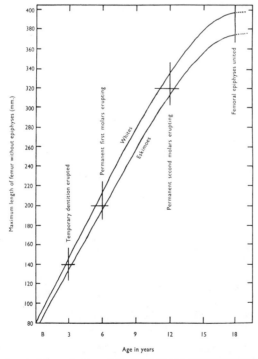

Fig. 111. Generalized postnatal growth curve of the femur (length of dried bone without epiphyses) based on a series of Eskimoes, the ages of which were estimated from the teeth. The accompanying generalized curve for Whites takes into account the known differences in size at maturity. The approximate age of a femur belonging to this size range can be determined by locating its length on the appropriate curve and noting the age listed directly below. From Stewart, 1968, Fig. 54, p. 133.

Permission for reproduction granted by Dr. T. Dale Stewart and John Wright and Sons, Ltd., Bristol.

Stewart (1954:407-450) has studied the growth of the femur from birth to adulthood and the following table presents data on the length of the femur without epiphyses compared with age (Fig. 111).

The epiphyses of the hip (head of femur) unite about 18 years while those of the knee fuse about 20 years of age. (See subadult section for closure of epiphyses).

The following tables by Anderson, Messner and Green (1964:1197-1202) (Tables 25-26) give femur length for children 1-18 taken from roentgenograms. Measurements were made of the entire bone, including both proximal and distal epiphyses. Caution should be used when dealing with archaeological material where one or both epiphyses may be missing and where the cartilaginous growth plate between the epiphysis and diaphysis is almost always missing. Anderson, Messner and Green state that measurement of the "femur was recorded as the distance from the proximal articulating surface of the capital epiphysis to the most distal point on the lateral condyle" (1964:1202). Bone lengths are in centimeters.

TABLE 25. BOYS: LENGTHS OF THE LONG BONES INCLUDING EPIPHYSES ORTHOROENTGENOGRAPHIC MEASUREMENTS FROM LONGITUDINAL SERIES OF SIXTY-SEVEN CHILDREN. AFTER ANDERSON, MESSNER AND GREEN'S TABLE 1 (1964:1198)

Femur

| No. | Age | Mean | σ_d | σ_m | Distribution | | | |
					$+2\sigma_d$	$+1\sigma_d$	$-1\sigma_d$	$-2\sigma_d$
21	1	14.48	0.628	0.077	15.74	15.11	13.85	13.22
57	2	18.15	0.874	0.107	19.90	19.02	17.28	16.40
65	3	21.09	1.031	0.126	23.15	22.12	20.06	19.03
66	4	23.65	1.197	0.146	26.04	24.85	22.45	21.26
66	5	25.92	1.342	0.164	28.60	27.26	24.58	23.24
67	6	28.09	1.506	0.184	31.10	29.60	26.58	25.08
67	7	30.25	1.682	0.205	33.61	31.93	28.57	26.89
67	8	32.28	1.807	0.221	35.89	34.09	30.47	28.67
67	9	34.36	1.933	0.236	38.23	36.29	32.43	30.49
6/	10	36.29	2.057	0.251	40.40	38.35	34.23	32.18
67	11	38.16	2.237	0.276	42.63	40.40	35.92	33.69
67	12	40.12	2.447	0.299	45.01	42.57	37.67	35.23
67	13	42.17	2.765	0.338	47.70	44.95	39.40	36.64
67	14	44.18	2.809	0.343	49.80	46.99	41.37	38.56
67	15	45.69	2.512	0.307	50.71	48.20	43.19	40.67
67	16	46.66	2.244	0.274	51.15	48.90	44.42	42.17
67	17	47.07	2.051	0.251	51.17	49.12	45.02	42.97
67	18	47.23	1.958	0.239	51.15	49.19	45.27	43.31

Femur

No.	Age	Mean	σ_d	σ_m	Distribution			
					$+2\sigma_d$	$+1\sigma_d$	$-1\sigma_d$	$-2\sigma_d$
30	1	14.81	0.673	0.082	16.16	15.48	14.14	13.46
52	2	18.23	0.888	0.109	20.01	19.12	17.34	16.45
63	3	21.29	1.100	0.134	23.49	22.39	20.19	19.09
66	4	23.92	1.339	0.164	26.60	25.26	22.58	21.24
66	5	26.32	1.437	0.176	29.19	27.76	24.88	23.45
66	6	28.52	1.616	0.197	31.75	30.14	26.90	25.29
67	7	30.60	1.827	0.223	34.25	32.43	28.77	26.95
67	8	32.72	1.936	0.236	36.59	34.66	30.78	28.85
67	9	34.71	2.117	0.259	38.94	36.83	32.59	30.48
67	10	36.72	2.300	0.281	41.32	39.02	34.42	32.12
67	11	38.81	2.468	0.302	43.75	41.28	36.34	33.87
67	12	40.74	2.507	0.306	45.75	43.25	38.23	35.73
67	13	42.31	2.428	0.310	47.17	44.74	39.88	37.45
67	14	43.14	2.269	0.277	47.68	45.41	40.87	38.60
67	15	43.47	2.197	0.277	47.86	45.67	41.27	39.08
67	16	43.58	2.193	0.268	47.97	45.77	41.39	39.19
67	17	43.60	2.192	0.268	47.98	45.79	41.41	39.22
67	18	43.63	2.195	0.269	48.02	45.82	41.44	39.24

Sex estimation:

As stated before, the femur is one of the most studied bones of the skeleton, and as such has contributed to the literature on sex determination.

The following data on the diameter of the femoral head were originally suggested by Pearson (1919:56) (Table 27). (See also Krogman 1962:144).

TABLE 27. RULES FOR SEXING THE FEMUR
AFTER PEARSON 1919: 56)

	FEMALE (MM.)	FEMALE? (MM.)	SEX? (MM.)	MALE? (MM.)	MALE (MM.)
Vertical diameter of head	<41.5	41.5– 43.5	43.5– 44.5	44.5– 45.5	>45.5
Popliteal length	<106	106 –114.5	114.5–132	132 –145	>145
Bicondylar width	<72	72 – 74	74 – 76	76 – 78	>78
Trochanteric oblique length	<390	390 –405	405 –430	430 –450	>450

It should be cautioned that Pearson's measurements were taken on 17th century London bones and that most modern populations are larger (Fig. 112).

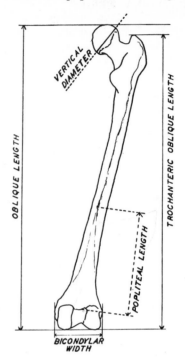

Fig. 112. Diagram of right femur as seen from behind, showing measurements referred to above. (After Pearson, 1917, Atlas plates 1 and 2.)

Thieme (1957) in discussing sex of the Negro skeleton gives the following data for femur length and femur head diameter (Table 28).

TABLE 28. AFTER THIEME'S TABLE 1 (1957:73)

Measurement	Sex	Number	Mean (mm.)	Standard Deviation	Standard Error of Mean	Critical Ratio ('t')
Femur length ..	M	98	477·34	28·37	2·866	10·13
	F	100	439·10	24·55	2·456	
Femur head diameter	M	98	47·17	2·75	0·278	16·17
	F	100	41·52	2·12	0·212	

Stature calculations:

The estimation of stature from long bones has been attempted by numerous authors. Only those formulas given by Trotter and Gleser (1952, 1958) will be given.

Stature formula for femur - male

White	Negro
2.32 Femur + 65.53 ± 3.94	2.10 Femur + 72.22 ± 3.91
Mongoloid	Mexican
2.15 Femur + 72.57 ± 3.80	2.44 Femur + 58.67 ± 2.99

Length of femur must be in centimeters.

Example for femur of white male that is 45.0 cm. (450 mm.).

2.32 (45.0) + 65.53 ± 3.94
104.40 + 65.53 ± 3.94

169.93 cm. - mean
Range 169.93 - 3.94 = 165.09 cm. Low
169.93 + 3.94 = 173.87 cm. High

TABLE 29. AFTER TROTTER AND GLESER TABLE 18 (1952: 495).

WHITE FEMALES		NEGRO FEMALES	
3.36 Hum + 57.97	± 4.45	3.08 Hum + 64.67	± 4.25
4.74 Rad + 54.93	± 4.24	2.75 Rad + 94.51	± 5.05
4.27 Ulna + 57.76	± 4.30	3.31 Ulna + 75.38	± 4.83
2.47 Fem_m + 54.10	± 3.72	2.28 Fem_m + 59.76	± 3.41
2.90 Tib_m + 61.53	± 3.66	2.45 Tib_m + 72.65	± 3.70
2.93 Fib + 59.61	± 3.57	2.49 Fib + 70.90	± 3.80
1.39(Fem_m + Tib_m) + 53.20	± 3.55	1.26(Fem_m + Tib_m) + 59.72	± 3.28
1.48 Fem_m + 1.28 Tib_m + 53.07	± 3.55	1.53 Fem_m + 0.96 Tib_m + 58.54	± 3.23
1.35 Hum + 1.95 Tib_m + 52.77	± 3.67	1.08 Hum + 1.79 Tib_m + 62.80	± 3.58
0.68 Hum + 1.17 Fem_m + 1.15 Tib_m + 50.12 [1]	± 3.51	0.44 Hum — 0.20 Rad + 1.46 Fem_m + 0.86 Tib_m + 56.33	± 3.22

[1] To estimate stature of older individuals subtract .06 (age in years – 30) cm;

Race estimation:

Stewart (1962:49-62) has studied anterior femoral curvature for its utility in race identification. The following table is reproduced from his data. Because the measurements he uses are not the standard anthropometric measurements, parts of his section on measurements are reproduced here.

METHODS

"A simple procedure was used to obtain the few measurements desired initially. Each femur was placed horizontally on a smooth table top so that it rested firmly on the posterior surfaces of the condyles at the distal end and of the greater trochanter (or perhaps better, the quadratus tubercle of the intertrochanteric ridge) at the proximal end. A wooden wedge was then inserted under the quadratus tubercle so as to raise the deepest point (bottom) of the anterior concavity at the proximal end of the shaft to the same level as the bottom of the anterior concavity at the distal end.* The proximal leveling point is located usually toward the medial side of the shaft and, when torsion is minimal, is just above the lesser trochanter, but as torsion increases the point moves distalwards. The distal leveling point is always toward the lateral side of the shaft and 1-2 cm proximal to the anterior margin of the lateral condyle.

Leveling was done by an improvised perigraph consisting of a triangular piece of board with a bisecting groove on the upper surface just large enough to hold firmly the fixed branch of a sliding caliper. In order to measure the height above the table surface of any particular point on the bone, the sliding branch of the caliper was brought into contact with the point, the reading made, and a factor added representing the height of the fixed branch of the caliper above the table surface.

With the bone in the described position, and using the "perigraph" as indicated, the following heights above the table surface were obtained: 1) at the leveling points, 2) at the point of greatest anterior curvature of the diaphysis, 3) at the highest point on the cervical tubercle (located, as shown in figure 113, on the anterior surface of the greater trochanter at the base of the neck), and 4 at the highest point on the head. The difference between heights 1 and 2 gives, of course, a measure of curvature. The last two heights were taken in order to get an indication of the amount of torsion in the proximal part of the bone. In this case, as figure 113 demonstrates, the greater the difference in height between the two points, the greater the torsion; and conversely, the smaller the difference, the smaller the torsion. Such a measure of torsion is not commonly used, but seems adequate for the present purpose, and has the advantage of being more easily obtained than the angle of the neck.

In advance of positioning the bone the greatest length between the most proximal point on the greater trochanter and the most distal point on the lateral condyle was obtained with the osteometric board. This measurement was used in ratios with two of the

---

*A descriptive problem is created by the fact that the femur is in the horizontal position. Although this position leads to little, if any, confusion as regards the meaning of the anatomical terms "anterior," "posterior," "lateral," "medial," "proximal," and "distal," it does lead to confusion when common terms of direction are used, such as "upper," "lower," "above" and "below." Once the femur is moved from its standard anatomical position, terms like "proximal" and "upper" or "superior" no longer share a common directional meaning. This explanation may help make my meaning clear when I speak of concavities on the anterior surface of the horizontally placed femur as having deepest points or bottoms; obviously the concavities open upwards (not proximalwards). As another example, the anterior concavity located at the proximal end of the horizontal shaft is often just above (not proximal to) the lesser trochanter.

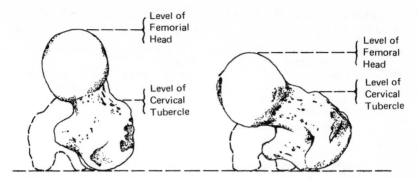

Fig. 113. Two femora with very different amounts of torsion viewed from a craniad position when the condyles and intertrochanteric ridges are in alignment. Note that the difference in level between the head and cervical tubercle constitutes an indicator of torsion. The actual figures obtained in these cases (Dakota Indians 325,354 and 325,369) were 39 mm and 19 mm, respectively.

heights. Later, when the bone had been leveled and the point of maximum diaphyseal curvature located, the distance between the proximal extremity of the greater trochanter and the point of greatest anterior curvature of the diaphysis was obtained with a large spreading caliper designed for use in pelvimetry. The ratio of these two length measurements was an important element in the study.

There is more than one way, of course, of measuring the length of the femur. The one used here relates to the shaft proper and ignores the neck and head. (Stewart 1962:49-62) (Table 30).

TABLE 30. SUMMARY OF FIRST SET OF MEASUREMENT (IN MM) AND INDICES OF FEMORA. AFTER STEWART'S TABLE 1 (1962:54).

MEASUREMENT OR INDEX	RACIAL GROUP	MEAN	S. D.	RANGE
1. Greater trochanter— lateral condyle length	Negro	450.6	24.1	411–500
	White	426.2	20.2	383–474
	Indian	433.3	18.9	404–482
2. Height of leveling points	Negro	61.3	2.9	55–67
	White	61.7	3.0	55–68
	Indian	64.4	2.6	59–72
3. Height of leveling points relative to length	Negro	*13.6*	*0.7*	*11.9–14.6*
	White	*14.5*	*0.8*	*13.3–16.4*
	Indian	*14.9*	*0.8*	*13.3–16.4*
4. Height of shaft above leveling points	Negro	7.6	2.4	4–15
	White	8.8	2.6	2–15
	Indian	10.9	3.0	6–20
5. Height of shaft relative to length (index of curvature)	Negro	*1.7*	*1.5*	*0.8–3.3*
	White	*2.1*	*0.5*	*0.5–3.5*
	Indian	*2.5*	*0.7*	*1.3–4.3*
6. Distance from greater trochanter to point of maximum curvature	Negro	204.6	28.9	159–287
	White	195.7	17.9	165–238
	Indian	230.3	27.9	185–308
7. Distance to point of maximum curvature relative to length (position index)	Negro	*45.3*	*5.5*	*35.9–62.1*
	White	*46.1*	*3.8*	*37.0–53.7*
	Indian	*53.2*	*6.3*	*43.4–70.3*
8. Indicator of torsion	Negro	15.9	6.4	0–28
	White	15.5	6.4	0–29
	Indian	25.3	5.1	15–39

PATELLA (KNEE CAP)

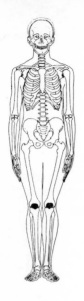

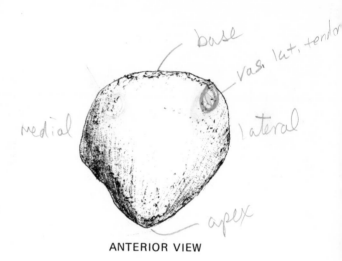

base

vas. lat. tendon

medial

lateral

apex

ANTERIOR VIEW

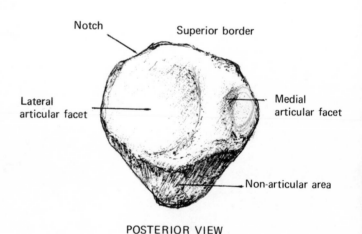

Notch

Superior border

Lateral
articular facet

Medial
articular facet

Non-articular area

POSTERIOR VIEW

Fig. 114.   LEFT PATELLA

Patella (KNEECAP): Paired—Irregular bone (Fig. 114)

Subadult bone:

The patella may ossify from several centers (Morris and Shaeffer 1953:258). Ossification usually occurs near 38 months (3-4 years) in males and 29 months

180

(2-3 years) in females (Pyle and Hoerr 1955:51) (Fig. 115). The ossification of the patella is usually complete by puberty or shortly thereafter.

PATELLA

Fig. 115.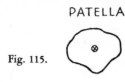

Adult bone:

A sesamoid (resembling a grain of sesame) is a flat bone which develops in a tendon that moves over a bony surface. Most sesamoid bones occur in the hands and feet where tendons cross the articulations for the metacarpals, metatarsals and phalanges, but the largest sesamoid bone in the body is the patella.

The kneecap is triangular in shape with its base proximal and its apex distal. The anterior side is marked by longitudinal striae and is slightly convex and perforated by small openings that transmit nutrient vessels.

The posterior is the articular surface and is divided by a slightly marked longitudinal surface corresponding to the groove on the patella surface of the femur. The larger lateral portion adapts to the lateral condyle and the smaller medial portion moves on the medial condyle.

Note that the superior border is thick and the inferior (or non-articular posterior) border is teardrop in shape or presents a blunt point.

BONES OF SIMILAR SHAPE WHERE CONFUSION MAY ARISE:

None.

Side identification:

1. Hold the nonarticular inferior blunt point between your fingers with the articular posterior surface toward you, or by the proximal border and the point as illustrated below. The side with the largest articular surface is on the same side the bone comes from (Fig. 115).

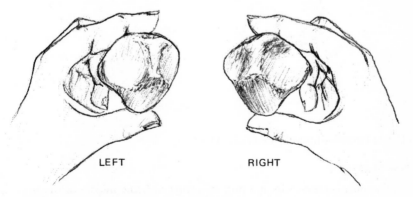

LEFT   RIGHT

Fig. 115. (cont.)   PATELLA

2. On many patellae there is a notch on the superior lateral border. When held as above, this notch is on the same side the bone is from.

TIBIA

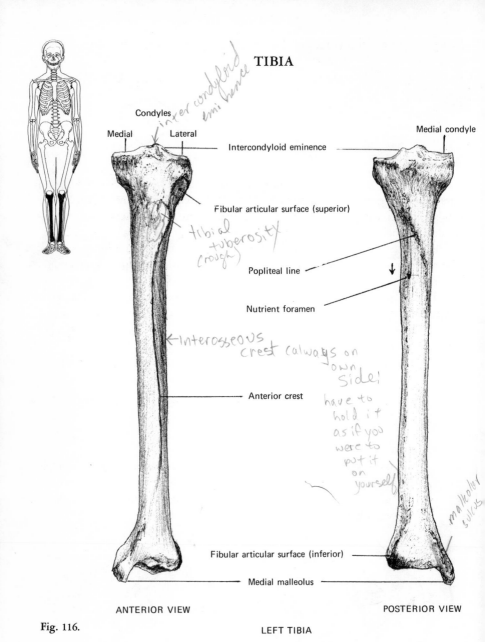

Condyles

Medial — Lateral

inter condyloid eminence *(handwritten)*

Intercondyloid eminence

Medial condyle

Fibular articular surface (superior)

tibial tuberosity (rough) *(handwritten)*

Popliteal line

Nutrient foramen

← Interosseous crest (always on own side! have to hold it as if you were to put it on yourself) *(handwritten)*

Anterior crest

malleolar sulcus *(handwritten)*

Fibular articular surface (inferior)

Medial malleolus

ANTERIOR VIEW

POSTERIOR VIEW

Fig. 116.

LEFT TIBIA

TIBIA: Paired—Long bone (Figs. 116, 117)

Subadult bone:

The tibia is ossified from a primary center near the middle of the shaft that appears about the 7th to 8th week of intra-uterine life and two centers for epiphyses (Fig. 118). The epiphysis at the proximal end (knee) appears at birth or slightly before and at the distal end during the first year (Hoerr and Pyle 1962).

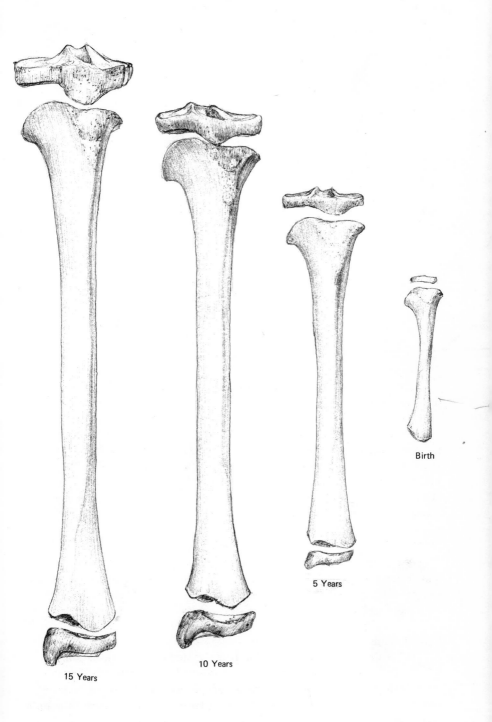

15 Years

10 Years

5 Years

Birth

Fig. 117.

Fig. 118.   OSSIFICATION OF THE TIBIA

The distal epiphysis usually unites to its shaft from age 13 (females) to 18 or some two years prior to fusion of the proximal end. McKern and Stewart (1957:49) found complete union for all of their American males by 20.

At the proximal epiphysis progress for complete union is slower and does not usually occur until the 23rd year. Pyle and Hoerr state that "the tibial epiphysis is the first epiphysis to fuse with its shaft at the knee" (1955:79) and they find beginning fusion at 14 in females and 16-17 in males. This again follows the general rule that complete fusion of the epiphyses at hip and ankle occurs about 18 years and at the knee about 20.

Adult bone:

The tibia is the largest bone in the lower leg and, after the femur, the second largest bone in the skeleton. The tibia or shin bone is situated on the anterior (front) and medial (inside) side of the leg. As with all long bones, it is divisible into a shaft and two extremities.

Superior end:
1. This consists of two superior articular surfaces for articulation with the femur. The articular surfaces are separated by the intercondyloid eminence, which is composed of the medial and lateral intercondyloid tubercles.
2. Just below the lateral femoral articular surface, i.e., on the inferior and lateral aspect of the lateral condyle, is the fibular articular surface, a flat and nearly circular area.

Shaft:

   1. The anterior surface of the shaft, containing the anterior crest, is easily
      felt below the skin of the living for more than half the length of the
      bone. Note that the bone can be felt on the medial side (inside) of the
      lower leg, and that when your fingers reach the crest along the anterior
      (front) of the leg as you proceed around or to the outside of the leg, the
      bone abruptly drops away.
   2. The posterior surface contains the nutrient foramen and the popliteal
      line and is slightly concave.

Distal end:

   1. The medial malleolus, which is easily felt on the medial side (inside) of
      the ankle is the projection at the distal end.
   2. A concavity occurs on the lateral surface opposite the medial malleolus
      for the articulation of the fibula. There is no lateral malleolus on the
      tibia.

BONES OF SIMILAR SHAPE WHERE CONFUSION MAY ARISE:

The tibia shaft could possibly be confused with the *femur shaft*. The tibia
shaft is triangular with sharp distinct edges, whereas the femur shaft is rounded
and smooth. The linea aspera found on the femur does not occur on the tibia.
The shaft of the *humerus* is smaller.

Side identification:

   1. When the bone is held in approximate anatomical position, the fibular
      articular surfaces are on the same side the bone is from. The medial mal-
      leolus is therefore opposite the side the bone comes from (Fig. 119a).

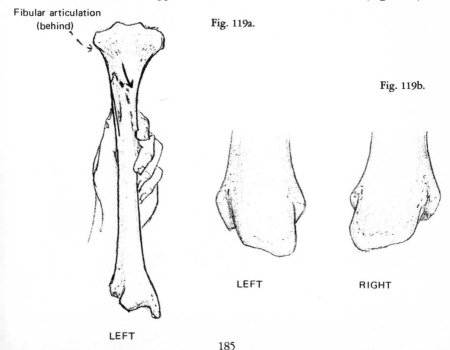

Fibular articulation
   (behind)

Fig. 119a.

Fig. 119b.

LEFT   RIGHT

LEFT

185

2. Hold the posterior surface toward you with the distal end away from you so that you can observe the nutrient foramen (inclined toward the distal end) and the popliteal line. The popliteal line is inclined toward the opposite side the bone comes from.
3. If you have just the distal end of the tibia, the tip of the medial malleolus is always toward the anterior surface of the bone when viewed from the medial side and will therefore be on the opposite side the bone comes from (Fig. 119b).
4. If you have only the shaft, locate the nutrient foramen and note that it is inclined distally and is in the posterior surface. Turn the bone over to the anterior surface and note that the sharp anterior crest abruptly drops away on the side the bone is from.

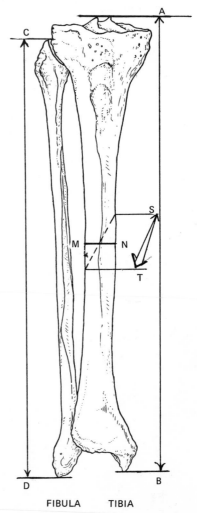

Fig. 120.   FIBULA   TIBIA

Measurements of the tibia (Fig. 120):

There are various slightly different methods for measuring the length of the tibia. Since this measurement is predominately used to calculate stature, I have given the technique employed by Trotter and Gleser (1952:473).

Maximum length: (osteometric board).

"End of malleolus against vertical [fixed] wall of the osteometric board, bone resting on its dorsal surface with its long axis parallel with the long axis of the board, block applied to the most prominent part of lateral half of lateral condyle" (Trotter and Gleser 1952: 473) (A-B).

Anterior-posterior diameter at nutrient foramen: (sliding caliper).

Maximum anterior-posterior diameter of shaft at level of nutrient foramen (at about the proximal one-third) (S-T).

Medio-lateral diameter at nutrient foramen: (sliding caliper).

Maximum transverse diameter at level of nutrient foramen (at right angles to the previous measurement) (M-N).

Platycnemic Index:

This expresses the degree of medio-lateral flatness of a tibia, from two dimensions taken at the level of the nutrient foramen.

$$\text{Platycnemic Index:} \quad \frac{\text{Medio-lateral nutrient diameter x 100}}{\text{Antero-posterior nutrient diameter}}$$

Range:

Hyperplatycnemic—X-54.9
Platycnemic—55.0-62.9
Mesocnemic—63.0-69.9
Eurycnemic—70.0-X
Neolithic bones from France—61.5-65.4*
Modern French—71.0-74.0
*From Wilder (1920: 131-132). See also Brothwell (1963: 92).

Fibula:

Maximum length: (osteometric board)

The maximum distance between the proximal and distal extremities. Follow same measuring procedure as with the humerus (C-D).

Age estimation:

Johnston (1962:249-254) has published on the growth of long bones from the infants and children of the Indian Knoll, Kentucky skeletal population. The relation of age to length of the subadult bone is as follows where bone length is in millimeters.

TABLE 31. AFTER JOHNSTON'S
TABLE 2 (1962:251)

Estimated age in years	Tibia		
	n	Mean	σ
Fetal	6	55.50	7.54
NB–0.5	65	69.28	6.33
0.5–1.5	38	96.87	14.47
1.5–2.5	7	120.57	5.45
2.5–3.5	10	138.20	8.54
3.5–4.5	10	154.30	8.10
4.5–5.5	7	178.43	4.53

The distal epiphysis unites before the proximal.

	Beginning union	Complete union
Distal epiphysis:	11-13 females	17
	14-16 males	20
Proximal epiphysis:	14 females	18
	16-17 males	23

The following tables by Anderson, Messner and Green (1964:1197-1202) give tibia length for children 1-18 taken from roentgenograms (Tables 32-33). Measurements were made of the entire bone including proximal and distal epiphyses. Caution should be used when dealing with archaeological material where one or both epiphyses may be missing and where the cartilaginous growth plate between the epiphysis and diaphysis is almost always missing. Anderson, Messner and Green state that measurement of the "tibia was recorded as the distance from the mid-point of a line drawn across the proximal condyles to the mid-point of the distal articulating surface" (1964:1202). Bone lengths are in centimeters.

TABLE 32. BOYS: LENGTHS OF THE LONG BONES INCLUDING
EPIPHYSES ORTHOROENTGENOGRAPHIC MEASUREMENTS FROM
LONGITUDINAL SERIES OF SIXTY-SEVEN CHILDREN. AFTER
ANDERSON, MESSNER AND GREEN'S TABLE 1 (1964:1198)

Tibia

| No. | Age | Mean | σ_d | σ_m | Distribution | | | |
					$+2\sigma_d$	$+1\sigma_d$	$-1\sigma_d$	$-2\sigma_d$
61	1	11.60	0.620	0.074	12.84	12.22	10.98	10.36
67	2	14.54	0.809	0.099	16.16	15.35	13.73	12.92
67	3	16.79	0.935	0.114	18.66	17.72	15.86	14.92
67	4	18.67	1.091	0.133	20.85	19.76	17.58	16.49
67	5	20.46	1.247	0.152	22.95	21.71	19.21	17.97
67	6	22.12	1.418	0.173	24.96	23.54	20.87	19.46
67	7	23.76	1.632	0.199	27.02	25.39	22.13	20.50
67	8	25.38	1.778	0.217	28.94	27.16	23.60	21.82
67	9	26.99	1.961	0.240	30.91	28.95	25.02	23.06
67	10	28.53	2.113	0.258	32.76	30.64	26.42	24.30
67	11	30.10	2.301	0.281	34.70	32.40	27.80	25.50
67	12	31.75	2.536	0.310	36.82	34.29	29.21	26.68
67	13	33.49	2.833	0.346	39.16	36.32	30.66	27.82
67	14	35.18	2.865	0.350	40.91	38.04	32.32	29.45
67	15	36.38	2.616	0.320	41.61	39.00	33.76	31.15
67	16	37.04	2.412	0.295	41.86	39.45	34.63	32.22
67	17	37.22	2.316	0.283	41.85	39.54	34.90	32.59
67	18	37.29	2.254	0.275	41.80	39.54	35.04	32.78

TABLE 33. GIRLS: LENGTHS OF THE LONG BONES INCLUDING
EPIPHYSES ORTHOROENTGENOGRAPHIC MEASUREMENTS FROM
LONGITUDINAL SERIES OF SIXTY-SEVEN CHILDREN. AFTER
ANDERSON, ET AL.
TABLE 2 (1964:1199)

Tibia

| No. | Age | Mean | σ_d | σ_m | Distribution | | | |
					$+2\sigma_d$	$+1\sigma_d$	$-1\sigma_d$	$-2\sigma_d$
61	1	11.57	0.646	0.082	12.86	12.22	10.92	10.28
67	2	14.51	0.739	0.090	15.99	15.25	13.77	13.03
67	3	16.81	0.893	0.109	18.60	17.70	15.92	15.02
67	4	18.86	1.144	0.140	21.15	20.00	17.72	16.57
67	5	20.77	1.300	0.159	23.37	22.07	19.47	18.17
67	6	22.53	1.458	0.178	25.45	23.99	21.07	19.61
67	7	24.22	1.640	0.200	27.50	25.86	22.58	20.94
67	8	25.89	1.786	0.218	29.46	27.68	24.10	22.32
67	9	27.56	1.993	0.243	31.55	29.55	25.57	23.57
67	10	29.28	2.193	0.259	33.67	31.47	27.09	24.89
67	11	31.00	2.384	0.291	35.77	33.38	28.62	26.23
67	12	32.61	2.424	0.296	37.46	35.03	30.19	27.76
67	13	33.83	2.374	0.290	38.58	36.20	31.46	29.08
67	14	34.43	2.228	0.272	38.89	36.66	32.20	29.97
67	15	34.59	2.173	0.265	38.94	36.76	32.42	30.24
67	16	34.63	2.151	0.263	38.93	36.78	32.48	30.33
67	17	34.65	2.158	0.264	38.97	36.81	32.49	30.33
67	18	34.65	2.161	0.264	38.97	36.81	32.49	30.33

Stature calculations:

The estimation of stature from long bones has been attempted by numerous authors. Only those formulas given by Trotter and Gleser (1952, 1958) will be given.

Stature formula for tibia - male

(See section on measurement for technique employed in measuring the tibia by Trotter and Gleser).

White	Negro
2.42 Tibia + 81.93 ± 4.00	2.19 Tibia + 85.36 ± 3.96

Mongoloid	Mexican
2.39 Tibia + 81.45 ± 3.27	2.36 Tibia + 80.62 ± 3.73

See Femur for example of how to use formula if necessary.

TABLE 34. AFTER TROTTER AND GLESER'S TABLE 18 (1952: 495)

WHITE FEMALES		NEGRO FEMALES	
3.36 Hum + 57.97	± 4.45	3.08 Hum + 64.67	± 4.25
4.74 Rad + 54.93	± 4.24	2.75 Rad + 94.51	± 5.05
4.27 Ulna + 57.76	± 4.30	3.31 Ulna + 75.38	± 4.83
2.47 Fem$_m$ + 54.10	± 3.72	2.28 Fem$_m$ + 59.76	± 3.41
2.90 Tib$_m$ + 61.53	± 3.66	2.45 Tib$_m$ + 72.65	± 3.70
2.93 Fib + 59.61	± 3.57	2.49 Fib + 70.90	± 3.80
1.39(Fem$_m$ + Tib$_m$) + 53.20	± 3.55	1.26(Fem$_m$ + Tib$_m$) + 59.72	± 3.28
1.48 Fem$_m$ + 1.28 Tib$_m$ + 53.07	± 3.55	1.53 Fem$_m$ + 0.96 Tib$_m$ + 58.54	± 3.23
1.35 Hum + 1.95 Tib$_m$ + 52.77	± 3.67	1.08 Hum + 1.79 Tib$_m$ + 62.80	± 3.58
0.68 Hum + 1.17 Fem$_m$ + 1.15 Tib$_m$ + 50.12 [1]	± 3.51	0.44 Hum — 0.20 Rad + 1.46 Fem$_m$ + 0.86 Tib$_m$ + 56.33	± 3.22

[1] To estimate stature of older individuals subtract .06 (age in years--30) cm;

FIBULA

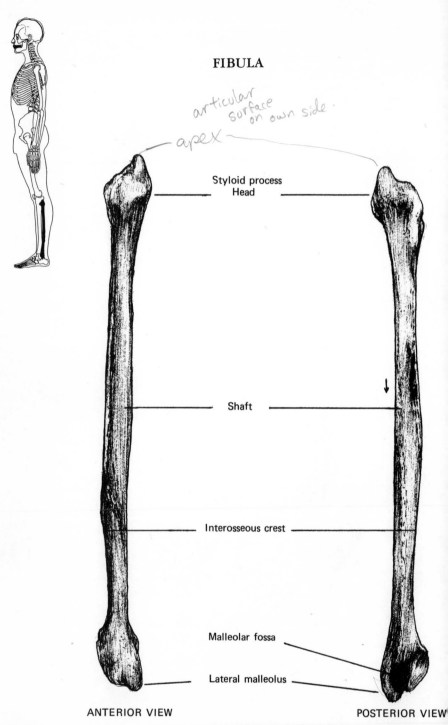

articular surface on own side.
apex

Styloid process
Head

Shaft

Interosseous crest

Malleolar fossa

Lateral malleolus

ANTERIOR VIEW

POSTERIOR VIEW

Fig. 121.

LEFT FIBULA

192

Birth

5 Years

10 Years

15 Years

Fig. 122.

193

FIBULA: Paired—Long Bone (Figs. 121, 122)

Subadult bone:

The fibula ossifies from a single center appearing about the 8th week of intra-uterine life near the center of the shaft (Fig. 123). The nucleus of the epiphysis for the distal end appears around age 1 (Hoerr and Pyle 1962:74).

Fig. 123.

OSSIFICATION OF THE FIBULA

The nucleus for the proximal epiphysis appears about 3 years in girls and 4 years in boys (Pyle and Hoerr 1955:52).

As is true in the tibia the distal epiphysis unites earlier than proximal by some two years.

The distal epiphysis unites at 11-12 in females and 14-15 in males (Hoerr and Pyle 1962:112-114). McKern and Stewart (1957:51) state that complete union occurred by the age of 20 in all of their cases.

The proximal epiphysis of the fibula unites after the proximal epiphysis of the tibia. It begins to unite between 14-15 in females and 16-17 in males (Pyle and Hoerr 1955:78-80). McKern and Stewart (1957:51) found that complete union does not occur until the 22nd year in males.

Adult bone:

Situated on the lateral side of the lower leg, the fibula is, in proportion to its length, the most slender of all the long bones. It articulates with the tibia proximally (at the knee) and with the tibia and talus distally (at the ankle). In amphibia the fibula is as large as the tibia where it articulates with the femur. It is smaller and bears less weight in reptiles, is still smaller in monotremes and marsupials, and has either disappeared or is represented by a fibrous band in horses and ruminants (for example cows, oxen, sheep, goats, deer and antelopes). In carnivora and primates the complete fibula exists but does not bear weight, and only in man does the fibular malleolus descend below the level of the tibial malleolus. In man it is important for muscle attachments and in the formation of the ankle joint.

The head of the fibula is a rounded expansion and medially has a circular articular facet. The distal end (lateral malleolus) is pyramidal in form, somewhat flattened from side to side, and forms the outside of the ankle joint.

BONES OF SIMILAR SHAPE WHERE CONFUSION MAY ARISE:

Radius and *ulna* because of the comparable size of the shafts. The *radius* shaft is triangular with a prominent interosseous crest. The surface opposite this crest (lateral) is thick and rounded, and the triangular edges are not prominent. The ulna also has a triangular shaft with an interosseous crest, but the surface opposite the interosseous crest (medial) has sharper, more distinct edges. The *fibula* shaft tends to be irregular but more closely resembles the shaft of the ulna than it does the radius. The nutrient foramen in the ulna is larger and more prominent than in the fibula.

Side identification:
1. The malleolar fossa, the deep depression behind and below the distal articular surface, always goes down (distally) and is in the back (posterior) (Fig. 124a).

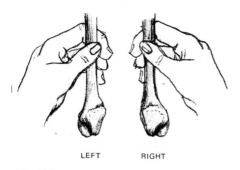

LEFT   RIGHT

Fig. 124a.

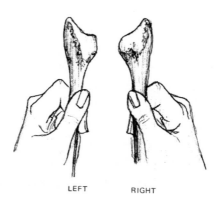

LEFT   RIGHT

Fig. 124b.

195

2. When held in anatomical position the articular facets on both the head and distal end are opposite the side the bone comes from.
3. If you have only the head (proximal end) hold the articular surface toward you and the styloid process will be on the same side the bone is from.
4. Hold the head up (superiorly) and the articular surfaces away from you; the twist of the distal half of the shaft is toward the side the bone is from (Fig. 124b).

Measurements of the fibula:

See section on tibia.

Age estimation:

Johnston (1962:249-254) has published on the growth of long bones from the infants and children of the Indian Knoll, Kentucky skeletal population (Table 35). The relation of age to length of the subadult bone is as follows where bone length is in millimeters.

TABLE 35. AFTER JOHNSTON'S
TABLE 2 (1962: 251)

Estimated age in years	Fibula		
	n	Mean	σ
Fetal	1	50.00	—
NB–0.5	37	65.38	5.19
0.5–1.5	25	92.44	13.79
1.5–2.5	5	113.80	7.44
2.5–3.5	6	134.17	10.39
3.5–4.5	7	144.71	11.32
4.5–5.5	3	171.67	4.52

The distal epiphysis unites before the proximal.

	Beginning union	Complete union
Distal epiphysis:	11–12 females	17
	14–15 males	20
Proximal epiphysis:	14–15 females	17
	16–17 males	22

Stature calculations:

The estimation of stature from long bones has been attempted by numerous authors. Only those formulas given by Trotter and Gleser (1952, 1958) will be given (Table 36).

Stature formula for fibula - male

White

2.60 Fibula + 75.50 ± 3.86

Negro

2.34 Fibula + 80.07 ± 4.02

Mongoloid

2.40 Fibula + 80.56 ± 3.24

Mexican

2.50 Fibula + 75.44 ± 3.52

See Femur for example of how to use formula if necessary.

TABLE 36. AFTER TROTTER AND GLESER'S TABLE 18 (1952:495)

WHITE FEMALES			NEGRO FEMALES	
3.36 Hum + 57.97		± 4.45	3.08 Hum + 64.67	± 4.25
4.74 Rad + 54.93		± 4.24	2.75 Rad + 94.51	± 5.05
4.27 Ulna + 57.76		± 4.30	3.31 Ulna + 75.38	± 4.83
2.47 Fem_m + 54.10		± 3.72	2.28 Fem_m + 59.76	± 3.41
2.90 Tib_m + 61.53		± 3.66	2.45 Tib_m + 72.65	± 3.70
2.93 Fib + 59.61		± 3.57	2.49 Fib + 70.90	± 3.80
1.39(Fem_m + Tib_m) + 53.20		± 3.55	1.26(Fem_m + Tib_m) + 59.72	± 3.28
1.48 Fem_m + 1.28 Tib_m			1.53 Fem_m + 0.96 Tib_m	
+ 53.07		± 3.55	+ 58.54	± 3.23
1.35 Hum + 1.95 Tib_m			1.08 Hum + 1.79 Tib_m	
+ 52.77		± 3.67	+ 62.80	± 3.58
0.68 Hum + 1.17 Fem_m			0.44 Hum — 0.20 Rad + 1.46	
+ 1.15 Tib_m + 50.12 [a]		± 3.51	Fem_m + 0.86 Tib_m + 56.33 ± 3.22	

To estimate stature of older individuals subtract .06 (age in years--30) cm.

FOOT

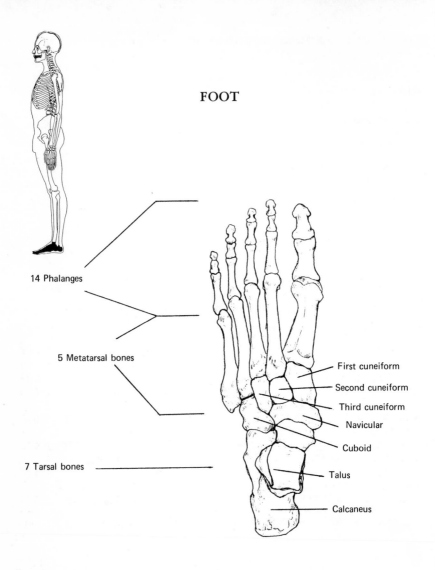

14 Phalanges

5 Metatarsal bones

First cuneiform

Second cuneiform

Third cuneiform

Navicular

Cuboid

7 Tarsal bones

Talus

Calcaneus

SKELETON OF LEFT FOOT (DORSAL VIEW)

Fig. 125.

Tarsal Bones: Paired—Irregular Bones (Fig. 125)

The tarsal bones are grouped in two rows; a proximal row consisting of two bones:

1. Talus
2. Calcaneus

And a distal row consisting of four bones:

    3. First cuneiform

    4. Second cuneiform

    5. Third cuneiform

    6. Cuboid

Interposed between the two rows on the tibial side (medial) of the foot is a single bone:

    7. Navicular

On the fibular side (lateral) the proximal and distal rows come in contact. Compared with the carpus, the tarsal bones present few common characteristics and a greater diversity of size and form.

Talus—The talus is the second-largest bone of the tarsus. Superiorly, it supports the tibia, and inferiorly it rests upon the calcaneus. At the sides it articulates with the two malleoli (tibia and fibula), and anteriorly it is thrust against the navicular (Fig. 126). No muscles are attached to it.

Side identification:

    Put the convex side up and head forward; the sharp angular side is on the side the bone is from.

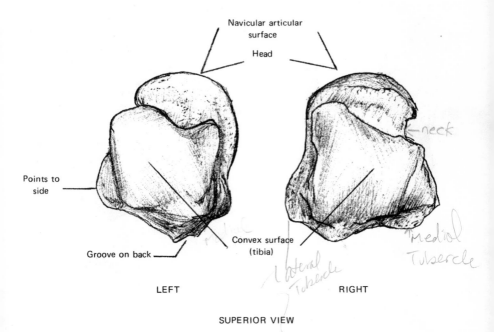

Fig. 126. Talus

Calcaneus—The calcaneus is the largest and strongest bone of the foot. It has an elongated form, is flattened from side to side and projects inferiorly and posteriorly to form the heel (Fig. 127).

Side identification:

Put heel toward you, and all articular surfaces are on top and on front of bone. Front point of bone is on the same side the bone comes from. Also the front facet (underneath point) is on the same side the bone comes from.

The shelf on the side of the bone is the sustentaculum tali which helps support the head of the talus. This shelf is opposite the side the bone comes from.

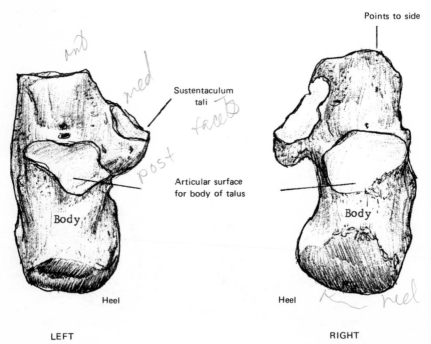

SUPERIOR VIEW

Points to side

Sustentaculum tali

Articular surface for body of talus

Body

Body

Heel

Heel

LEFT

RIGHT

Fig. 127.

200

Cuboid—The cuboid is on the lateral side (outside) of the foot and is in line with the calcaneus and 4th and 5th metatarsals (Fig. 128 a-e).

Side identification:

Place the pointed articular facet toward you, smooth nonfacet side up. The point of the articular facet will be opposite the side the bone comes from. The aspect of the bone with facet points to side from which it comes. There is also a slight groove on side from which it comes.

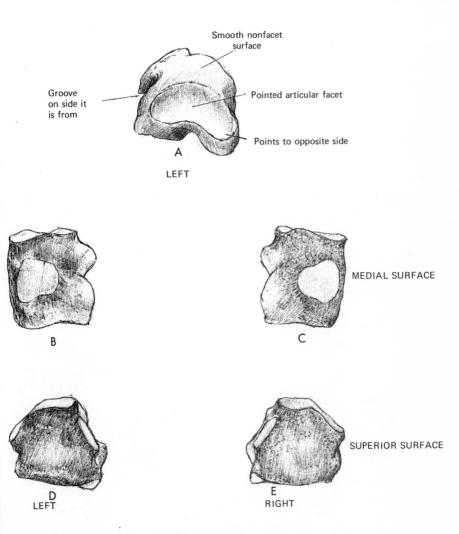

Fig. 128.

Cuneiform Bones—of the three cuneiform bones, the first is the largest, the second is the smallest, and the third intermediate in size. They are wedge-shaped bones.

First Cuneiform (Fig. 129a-c)

Side identification:

Largest of the three cuneiform bones. Place kidney-shaped articular facet (for first metatarsal) forward, away from you and the concave navicular facet toward you. The only other facet will be on top and the broad side will be down. The superior facet points to the side the bone comes from.

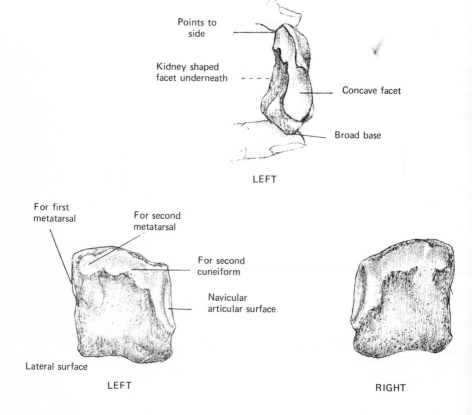

Fig. 129. First Cuneiform

Second Cuneiform (Fig. 130a-e)

Side identification:

Smallest of the three cuneiform bones. Place wedge (point) down, concave facet (navicular) toward you, and flat or smooth side up. On the side the bone is from there is an L-shaped facet (upside down as bone is now situated). This L-shaped facet is a complete facet. On the opposite side is an interrupted facet.

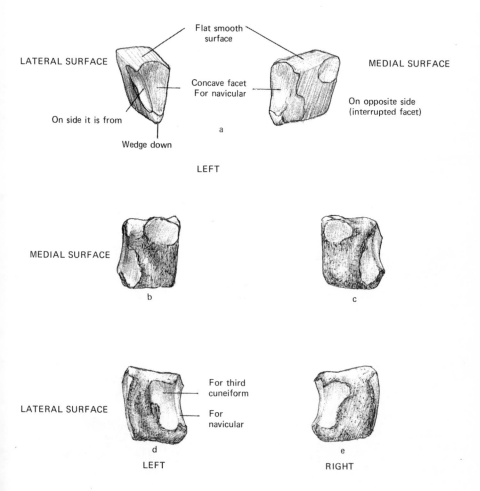

Fig. 130. Second Cuneiform

Third Cuneiform (Fig. 131a-e)

Side identification:

Intermediate in size between the first and second cuneiform. Place smooth broad nonarticular surface up, point or wedge down. Rotate in this position until the end closest to you and the one farthest away will both be articular surfaces. One of these articular surfaces (for navicular) rounds the edge of the bone and forms a small articular facet (for second cuneiform) that does not extend along the top of the entire side of the bone. This articular surface is opposite the side the bone comes from. The bone will also be slightly concave toward the opposite side the bone comes from.

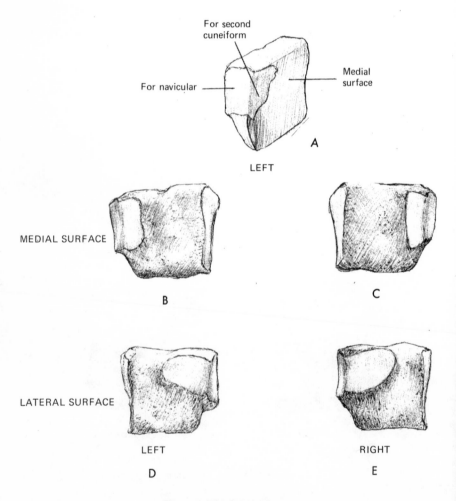

Fig. 131. Third Cuneiform

Navicular (Fig. 132a-c)—The navicular is characterized by a large oval, concave, articular facet on the posterior surface that receives the head of the talus. It sits on the inside of the foot. It is much larger than the navicular of the wrist and no confusion between the two should result.

Side identification:

Hold concave side toward you, three facets (underneath) away from you. Hold tubercle between thumb and forefinger and the bone inclines or points toward side it is from.

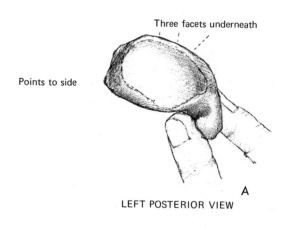

Three facets underneath

Points to side

A

LEFT POSTERIOR VIEW

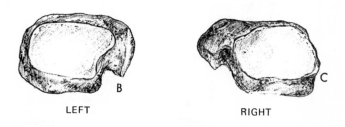

B

LEFT

C

RIGHT

Fig. 132. Navicular

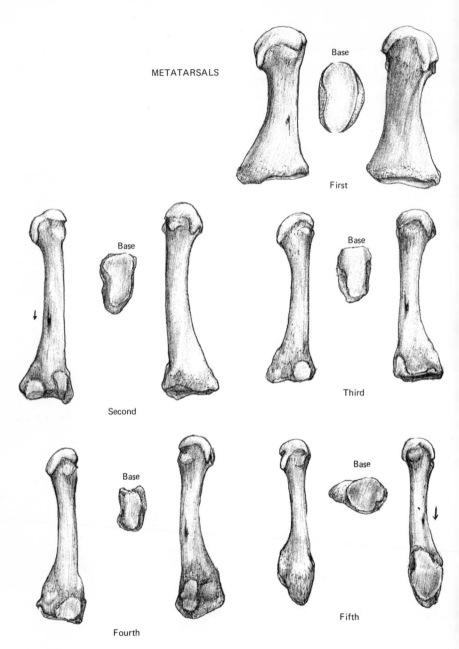

METATARSALS

Base

First

Base

Base

Second

Third

Base

Base

Fourth

Fifth

Fig. 133. Metatarsals (all are left bones with fibular, base, and tibia views)

Metatarsals: Paired—Short Bones (Fig. 133).

The metatarsals consist of five cylindrical bones in each foot. They articulate with the tarsus proximally and with the first or proximal row of phalanges distally. As with the metacarpals in the hand, they have been well described as long bones in miniature. As they extend from the tarsus they diverge slightly from each other. They are numbered from the medial (big toe) to the lateral side (small toe).

As with long bones, the metatarsals present a shaft and two extremities. The base or tarsal extremity articulates with the tarsus proximally, and on the side of the basilar end with the adjacent metatarsal bones.

The shaft tapers gradually from the base to the distal end (head) and is curved so as to be convex on the dorsal side (top of foot) and concave on the plantar side (bottom of foot). The shaft is somewhat triangular in cross section.

The head or distal extremity articulates with the proximal (first) phalange. Note that the heads all present large rounded articular surfaces. Although the nutrient foramina are difficult to find in many metatarsals, it should be noted that they are inclined toward the distal end of the first, but toward the proximal end of the second through the fifth (the same pattern as in the metacarpals).

First (big toe)—shortest, thickest and most massive of the metatarsals. The base has a saddle-shaped articular facet for the first cuneiform bone. The head is marked on the plantar surface by two deep grooves, separated by a ridge. These grooves are associated with two sesamoid bones.
Second—longest metatarsal.
Third, fourth and fifth successively decrease in length.

Side identification:
Anatomical position for all metatarsals will be the same when held with the proximal end (base) toward you.
First—the saddle-shaped articular surface will be slightly inclined toward the opposite side the bone comes from (Fig. 134a).

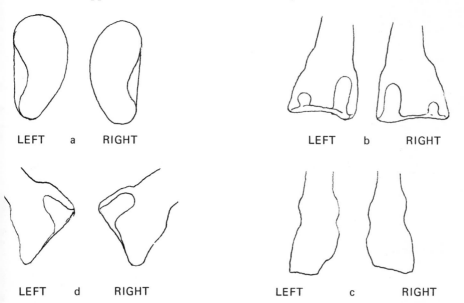

LEFT   a   RIGHT   LEFT   b   RIGHT

LEFT   d   RIGHT   LEFT   c   RIGHT

Fig. 134.

Second—usually the longest of the metatarsals, it has two small articular surfaces on the same side of the base the bone comes from (Fig. 134b).

Third—there is an articular surface that is continuous from the upper left side across the base and around to the upper right side (Fig. 134c). The largest articular facet on the side of base is on the same side the bone comes from. In some cases there may be a small articular surface along the lower margin opposite the side the bone comes from.

Fourth—usually shorter than the third, the base resembles that of the third very closely. When viewing the dorsal surface of the base, the base projects further on the same side the bone comes from (Fig. 134d).

Fifth—the fifth metatarsal is easily recognized by the rough nonarticular eminence known as the tuberosity on the lateral side of its base. With the groove between the articular facet and the tuberosity down, the tuberosity will be on the same side the bone comes from.

Phalanges (Toes): Paired—Short Bones (Fig. 135)

There are fourteen phalanges in each foot, two for the big toe and three for each of the other digits. They are divided into three rows:

First or proximal—5 (the largest).

Second or middle—4 (no middle phalanx in the big toe).

Third or distal or terminal—5; they are the smallest.

In all of the phalanges the nutrient canal (foramen) is directed toward the distal extremity. They are not easily seen. The phalanges present a shaft and two extremities.

First or proximal row (Fig. 135c):

The phalanges are constricted in the middle of the shaft and expand at either extremity. The proximal extremity presents a slightly concave, oval articular surface which receives the convex head of the metatarsal bone. The distal extremity is grooved in the center and elevated on each side into two little condyles. The two small condyles on the distal extremity of the proximal and middle phalanges resemble the distal or lower end of the femur. To correspond with these condyles, the bases of the terminal and middle phalanges have two small depressions, and resemble the proximal or upper end of the tibia.

Second or middle row (Fig. 135b):

These are stunted and much smaller bones than the corresponding phalanges of the fingers. The base or proximal extremity presents two shallow depressions separated by a medial ridge, which articulate with the first row of phalanges. The distal end articulates with the base of the third row of phalanges, and is grooved in the center and elevated on each side into two small condyles.

Third or distal row (Fig. 135a):

Small in size, the third phalanx is easily recognized because the distal end is tapered. One side (dorsal) is fairly smooth, over which the toenail fits, and the other (plantar) surface is rough because of the attachments of the fiber bands of the pulp of the digits.

The proximal end is similar in shape to that of the second phalanx in that it presents two shallow depressions separated by a medial ridge. One could never confuse the phalanges of the middle and distal row if it is remembered that both ends of the middle phalanges have articular surfaces, whereas only the proximal end is an articular surface on the phalanges in the distal row.

PHALANGES (TOES)

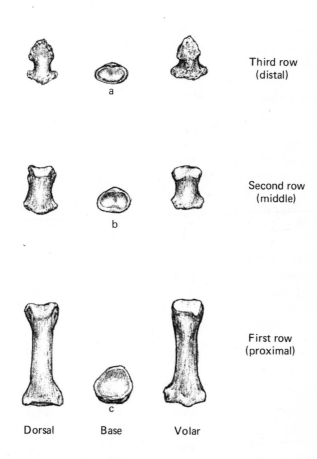

Third row
(distal)

a

Second row
(middle)

b

First row
(proximal)

Dorsal   Base   Volar

c

THE PHALANGES OF THE THIRD DIGIT OF THE FOOT
(Three views: dorsal, base and volar)

Fig. 135.

IV. HUMAN DENTITION*

HUMAN DENTITION
32 TEETH IN ADULTS
20 TEETH IN CHILDREN

The study of teeth is very important to the anthropologist and paleontologist. Because they are constructed of dense and hard material, teeth resist decay in the ground and often outlast bone. Consequently, teeth have played an important role in the study of fossil man.

The paleontologist spends much time studying the various genetic and functional characteristics of the teeth. The study of teeth is a special field of knowledge in itself; the anatomical terms used to describe the single tooth and the dental arch differ slightly from those used in anatomy.

There are four "types" of teeth in the human dental arch: incisors, canines, premolars and molars. This classification is based on both the morphology and function of the respective teeth. The actual genetic or developmental reasons for this morphology are unknown. However, it has been suggested that the dentition is under the influence of a morphogenetic field which controls its morphological expression. This theory has been advanced by P. M. Butler (1937, 39, 61, 63) and applied to the human dentition by A. Dahlberg (1945, 49, 63). Dahlberg feels that different tooth groups are under different morphogenetic fields and that these fields are concentrated as certain teeth within the tooth group. This produces the distinct "types" of teeth as well as some teeth that appear more stable and less variable than others within the group. Whatever the developmental reasons behind the formation of distinct types of teeth, they are usually expressed in a "dental formula" such as the following.

$$\text{ADULT DENTAL FORMULA} - I\frac{2}{2} \ C\frac{1}{1} \ PM\frac{2}{2} \ M\frac{3}{3} = 16 \times 2 = 32$$

Where

I = incisor—a tooth designed for cutting
C = canine—pointed cusp for tearing and incising
PM = premolars—broad occlusal surfaces with multiple cusps for
M = molars   grinding and reducing food material as an aid to digestion.

The dental formula expresses the number of teeth in the upper jaw (numbers above the line (1) and in the lower jaw (numbers below the line) but in only one-half of the mouth. The numbers recorded in the formula must be multiplied by 2 to determine the total number of teeth in the mouth. Often teeth will be given as M^1, M^2 etc., indicating the first and second upper molars and M_1, M_2 etc., indicating lower molars (Fig. 136a).

*The basis of this chapter was by Douglas H. Ubelaker

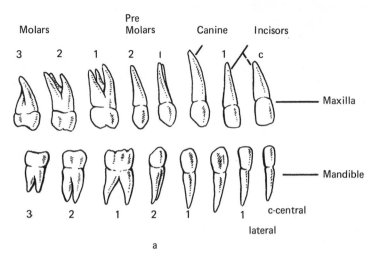

Molars | Pre Molars | Canine | Incisors

3   2   1   2   1   1   c

Maxilla

Mandible

3   2   1   2   1   1   c-central

lateral

a

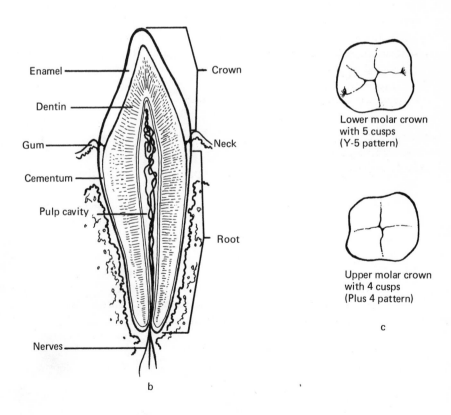

Enamel ——————— Crown

Dentin —————

Gum ———

Neck

Cementum ———

Pulp cavity ———

Root

Nerves ———

b

Lower molar crown
with 5 cusps
(Y-5 pattern)

Upper molar crown
with 4 cusps
(Plus 4 pattern)

c

Fig. 136. Human Dentition.

Man develops two sets of teeth. Usually at birth no teeth are visible, but at around six months the first deciduous (baby) teeth, the lower central incisors, erupt. The deciduous set consists of 20 teeth and may be given in a formula as follows:

$$\text{DECIDUOUS DENTAL FORMULA} - \text{i}\frac{2}{2} \ \text{c}\frac{1}{1} \ \text{m}\frac{2}{2} = 10 \times 2 = 20$$

Where the names of the teeth are abbreviated in the same way but with lower case letters. (Note that there are no premolars in the human deciduous formula.)

The deciduous dentition consists of 8 incisors, 4 canines and 8 deciduous molars. The deciduous incisors and canines are miniature replicas of the adult incisors and canines. The same criteria used to differentiate the adult incisors and canines may be employed on the corresponding deciduous teeth. The first deciduous molars in the maxillary are the precursors of the adult maxillary (upper) premolars. Similarly the first deciduous molars in the mandible correspond to the adult mandibular (lower) premolars. The deciduous second molars are replicas of the permanent first molars in the respective maxilla or mandible.

Sexual dimorphism is not marked enough in either the adult or deciduous dentition to allow sex determinations to be made. However, as a general rule, males tend to have slightly larger teeth.

Anatomical Terms For Teeth (Fig. 136b):

Every tooth has three areas:
1. Crown—that part of the tooth situated above the gum and covered with enamel.
2. Neck—a slightly constricted portion just below the crown and the area known as the cemento-enamel junction.
3. Root—that portion of the tooth below the crown and neck. It is enclosed in the tooth socket and covered with cementum.

Each tooth consists of:
1. Enamel—a white, compact and very hard substance that covers and protects the dentin of the crown of the tooth.
2. Cementum—a layer of bony tissue covering the root of a tooth.
3. Dentin—the chief tissue of the teeth which surrounds the pulp cavity. It is covered by enamel on most of the exposed parts of the tooth and by cementum on the part implanted in the jaw. Dentin forms the main bulk of the tooth.
4. Pulp cavity—the pulp chamber and canal within the tooth. It contains a soft tissue called pulp.

Each tooth has five surfaces (Fig. 137):
1. Labial (lips) or Buccal (cheek). Labial—the side toward the lips, is used with incisors and canines; buccal—the side toward the cheek, is used with premolars and molars.

2. Lingual—the side toward the tongue.
3. Occlusal—the surface of the tooth which comes into contact with the teeth of the opposite jaw (the biting surface).
4. Mesial—the surface of the tooth that lies against an adjoining tooth and faces toward the median line.
5. Distal—the surface of the tooth that lies against an adjoining tooth and faces away from the median line.

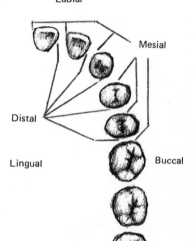

Labial

Mesial

Distal

Lingual

Buccal

Fig. 137. Surfaces of teeth

Occlusal

One of the best books for study of teeth is J. E. Anderson's *The Human Skeleton, A Manual for Archaeologists.* The serious student should consult this book for additional details.

When one is faced with the problem of identifying bones or teeth, it is most efficient to proceed by following certain steps (see steps for bones in the introduction). We recommend five steps as follows:

Step 1—Determine whether the tooth (teeth) is deciduous or adult. Deciduous teeth are (A) smaller than their adult equivalents and (B) are more yellow in color because the enamel and cementum layers are not as thick as in adult teeth.

Step 2—Is the tooth an incisor, canine, premolar (in adults) or molar? This is a fairly easy step.

Step 3—Is the tooth from the upper jaw (maxilla) or lower jaw (mandible)? This is a difficult step and requires study. Some students would rather reverse steps 3 and 4, and this can be done without any difficulty (see Anderson's four steps, which are not in the same sequence).

Step 4—What position in its group does it hold? Once the type of tooth has been determined, decide whether it is central or lateral incisor; a 1st or 2nd premolar, or a 1st, 2nd, or 3rd molar. This is a more difficult step and requires a considerable knowledge of dental anatomy.

Step 5—Is the tooth from the right or left side? This is probably the most difficult step, and comparative dental arches should be consulted.

A series of comparative dental arches, both upper and lower, should be consulted when identifying teeth, just as a skeleton should be consulted when identifying bones when there is some doubt in the identifier's mind.

Replacing Teeth in a Dental Arch:

When one is required to replace teeth in a dental arch, he must first know the various types of teeth and their locations. Of equal importance are wear facets located between the teeth (the mesial and distal edges). Teeth move as the jaw is in operation and therefore rub against one another, producing slick wear surfaces. These will match perfectly when the proper two adjacent teeth are placed together.

NEVER glue a tooth into a socket until the wear surfaces are matched (excessive wear on the occlusal surface due to attrition may occasionally eliminate this valuable aid.)

The following analysis presents data on each type of tooth and follows the steps outlined here.

INCISORS (Fig. 138)
4 CENTRAL—2 UPPER AND 2 LOWER
4 LATERALS—2 UPPER AND 2 LOWER

The incisors are the two teeth on either side of the midline in both jaws. They are characterized by single roots and crowns with a sharp occlusal (mesiodistal) ridge or edge.

They are the most frequently lost teeth in archaeological specimens, and the most frequently encountered outside of the dental arch as they have a short single root.

Step 1—Is it deciduous or adult?

Deciduous incisors are considerably smaller than adult and are more yellow in color. If the dental arch is present (either upper or lower) there will be spaces between the teeth as the individual approaches age six. The roots of teeth are formed only after growth of the crown is completed. (Once the roots of teeth have formed the teeth do not get larger, but as the child grows from age 2 to 6 the mandible and maxilla enlarge and create spaces between the teeth).

Step 2—What type of tooth is it?

An incisor has:

1. A single root (not as large as a canine).
2. A single crown with an occlusal (mesio-distal) edge.

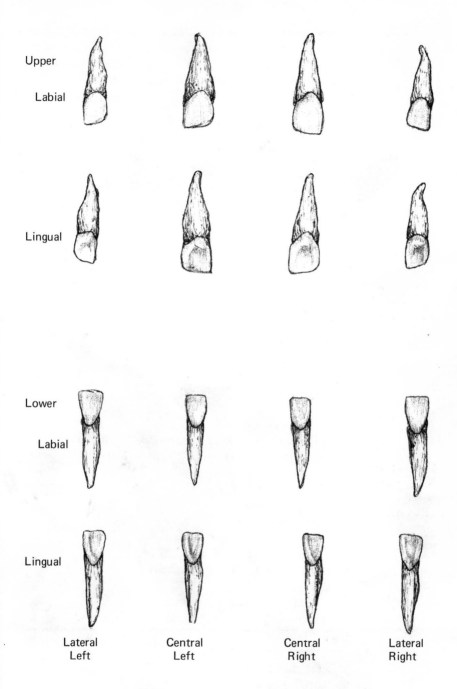

Upper

Labial

Lingual

Lower

Labial

Lingual

Lateral
Left

Central
Left

Central
Right

Lateral
Right

Fig. 138. Incisors

3. A shovel-shaped lingual surface, particularly in Mongoloid racial groups. In some cases an enamel extension is also present on the labial surface producing a "double-shovel shaped" incisor. Occasionally the lingual extension will enclose an area in the center and produce a "barrel-shaped" incisor (Fig. 139).

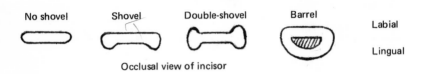

Occlusal view of incisor

Fig. 139.

Step 3—Is it upper or lower?
    An upper incisor:
        1. Is large
        2. Has a shovel-like crown
        3. Has a cingulum
    A lower incisor:
        1. Is small
        2. Has a narrow crown
        3. Has no cingulum

The cingulum is a bulge or raised area on the lingual surface of the tooth near its neck or gum line. It is usually absent from mandibular (lower) incisors.

Step 4—What position does it hold?
    An upper central incisor:
        1. Is the largest of all incisors.
        2. Has a square mesial angle of crown.
        3. Has a rounded distal angle of crown.
        4. Is most likely to have a shovel shape.
    An upper lateral incisor:
        1. Is smaller than an upper central.
        2. Usually has a pit at the base of the cingulum.
        3. May have shovel shaping.
    A lower central incisor:
        1. Is the smallest of all incisors.
    A lower lateral incisor:
        1. Is larger than a lower central but smaller than an upper.
        2. Has a wider crown (spreading out into a fan shape) at the occlusal surface.
Always be aware of the wear facets between neighboring teeth.

Step 5—Is it right or left?

Upper:

    1. The angle formed by the mesial and occlusal edges is a right angle (Fig. 140a).

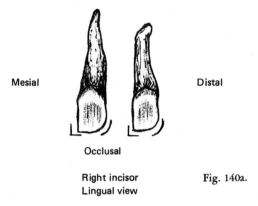

Mesial   Distal

Occlusal

Right incisor   Fig. 140a.
Lingual view

    2. The angle formed by the distal and occlusal edges is rounded.

Lower:

    The roots of lower incisors are flattened in a plane that is perpendicular to the axis of the crown. The roots are wider labio-lingually than mesio-distally. When held in proper position (by root with lingual surface facing you) there will be a groove on the flat surface of the root (distal surface) on the same side the tooth comes from (Fig. 140b).

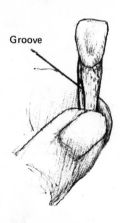

Groove

Fig. 140b.   Left incisor
Lingual view

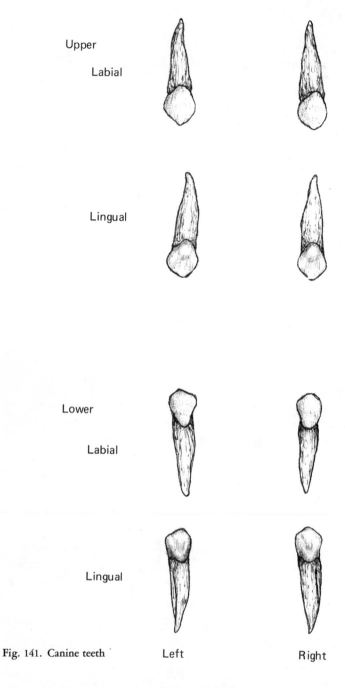

Upper

Labial

Lingual

Lower

Labial

Lingual

Fig. 141. Canine teeth

Left

Right

CANINE (Fig. 141)
2 UPPERS—1 RIGHT AND 1 LEFT
2 LOWERS—1 RIGHT AND 1 LEFT

The canines are sometimes called eye teeth (because of their position below the eyes) or dog teeth (because of their large size in that animal). This is the tooth which gave the Saber-toothed tiger its name. Although greatly reduced in man, it is still the longest tooth in the mouth and has the largest root in relation to crown surface of any tooth.

These are the second most frequently lost teeth in archaeological specimens or the second most frequently encountered teeth outside of the dental arch because they have one long single root.

Step 1—Is it deciduous or adult?

A deciduous canine is smaller than an adult canine and more yellow in color.

Step 2—What type of tooth is it?

A canine has:

1. A single large root (larger than an incisor).
2. A single pointed cusp.
3. A large root in relation to its crown.

Step 3—Is it upper or lower?

An upper canine has:

1. A wider crown.
2. A larger size.
3. A sharper single point cusp.
4. A cingulum (see incisor section for definition).

A lower canine has:

1. A narrower crown.
2. A smaller size
3. A blunt single pointed cusp.
4. No cingulum.

Step 4—What position does it hold?

This does not apply to canines since there is only one.

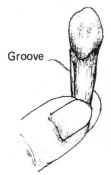

Groove

Left canine
Lingual view

Fig. 142a.

Left canine
Lingual view

Fig. 142b.

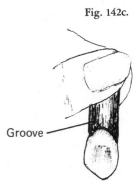

Fig. 142c.

Groove

Left canine
Lingual view

Step 5—Is it right or left?

Lower:

When held in proper position (by root with lingual surface facing you) there will be a groove on the flat surface of the root (distal surface) on the same side the tooth comes from (Fig. 142a).

If the occlusal surface is not worn down and can be observed it will be noted that the mesial slope of the crown is shorter than the distal slope. Viewed from the lingual surface the longer distal slope will be on the same side the tooth comes from (Fig. 142b).

The gritty diet of many aboriginal populations often produced so much wear on the occlusal surfaces of the crowns that these criteria may not be accurately applied.

Upper:

When held by root with crown pointing down and lingual surface facing you, there will be a groove on the flat surface of the root (distal surface) on the same side the tooth is from (Fig. 142c).

Always be aware of the wear facets between neighboring teeth.

PREMOLARS
4 UPPER—2 RIGHT AND 2 LEFT
4 LOWER—2 RIGHT AND 2 LEFT (Fig. 143)

The premolars, sometimes known as bicuspids because they usually have two cusps or points on the crown, may in man have one, two or three cusps. They are the two teeth behind the canines and in front of the molars.

These are not lost from archaeological specimens as often as are incisors and canines because their roots tend to be more complex, holding the tooth in its socket. Lower teeth are easier to remove as they have single roots.

Step 1—Is it deciduous or adult?

It has to be adult because there are no deciduous premolars.

Step 2—What type of tooth is it?

A premolar:

1. Usually has two cusps, one buccal and one lingual.

2. Is usually smaller than a molar.

Step 3—Is it upper or lower? *(Be cautious with worn teeth)*

An upper premolar:

1. Has cusps of equal size.

2. Usually has two roots; one buccal and one lingual (same as cusps).

3. May have fused roots, but the line of union can be seen.

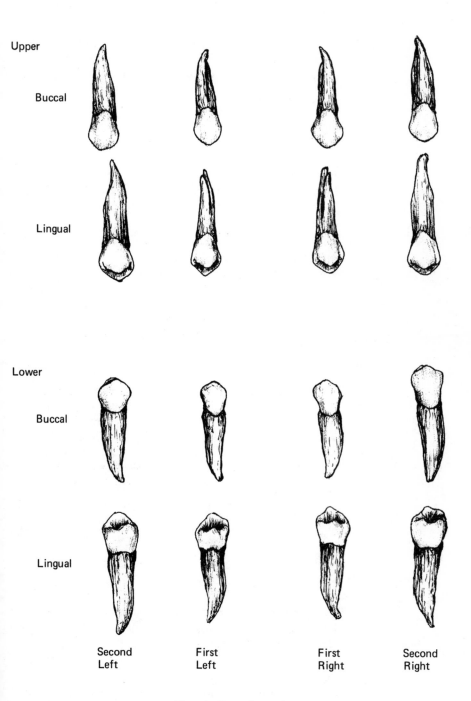

Upper

Buccal

Lingual

Lower

Buccal

Lingual

| Second Left | First Left | First Right | Second Right |

Fig. 143. Premolar teeth

221

A lower premolar has:

1. A large buccal cusp.
2. A single root, wider bucco-lingually and narrower mesio-distally.
3. A root tip that curves distally when viewed from the lingual surface.

Step 4—What position does it hold?

Upper Premolars (Fig. 144a)

First

1. Usually has 2 roots.
2. Buccal cusp may be larger.
3. Mesial surface of crown is concave.

Second

1. Usually has 1 root.
2. Both cusps are about equal.
3. Mesial surface of crown is convex.

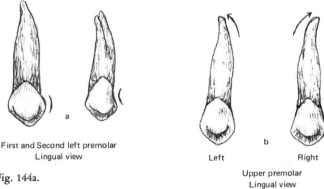

First and Second left premolar
Lingual view

Fig. 144a.

b

Left   Right

Upper premolar
Lingual view

Fig. 144b.

Lower Premolars

First

1. Has a small single lingual cusp.
2. May have a groove on the mesial surface of its root.
3. May have a larger buccal cusp.

Second

1. Has a small, sometimes double lingual cusp.
2. Has no groove on the mesial surface of its root.
3. Has cusps of equal size.

Step 5—Is it right or left?

Upper (Fig. 144b):

When held by root with crown pointing down and lingual surface facing you, the tip of the root will incline toward the same side the tooth comes from.

Lower (Fig. 144c):

When held by root with lingual surface facing you the tip of the root will incline toward the same side the tooth comes from.

First lower premolar may have a groove on mesial surface of root or on the root surface opposite the side the tooth is from (when held by root with lingual surface facing you).

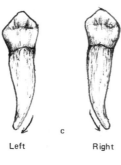

Left   Right

Lower premolar
Lingual view

Fig. 144c.

MOLARS
6 UPPER—3 RIGHT AND 3 LEFT
6 LOWER—3 RIGHT AND 3 LEFT (Fig. 145)

Most skulls will have three molars but many may have only two. The third molar is a genetically unstable tooth that may be lost in many individuals, thus reducing the total number of adult teeth from 32 to 28. The molars are the grinding teeth that make up the dental arch behind the premolars. They are fair age indicators, as the first erupts about 6 years (6 year molar), the second at about 12 years (12 year molar), and the third at 18 years (wisdom tooth). The third molar, being genetically unstable, may erupt at any age from 18 years to time of death.

These are the least frequently lost teeth in archaeological specimens because of their multiple roots. Indeed, they are difficult to remove from their tooth sockets.

Step 1—Is it deciduous or adult?

Deciduous molars:

1. Are much smaller.
2. Are more yellow in color.
3. Have much thinner roots.
4. Have roots that are much wider apart.

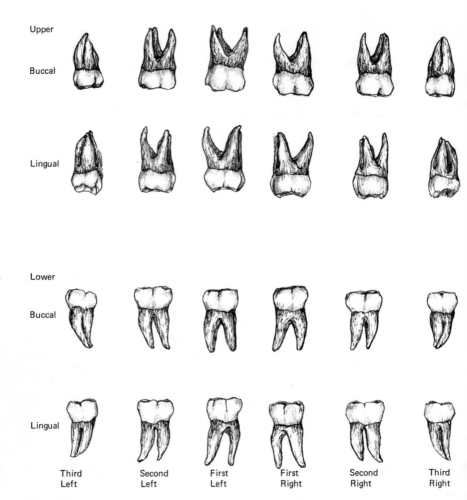

Upper						
Buccal						
Lingual						
Lower						
Buccal						
Lingual						

| Third
Left | Second
Left | First
Left | First
Right | Second
Right | Third
Right |

Fig. 145. Molar teeth

Step 2—What type of tooth is it?

    A molar has:

        1. Multiple cusps.

        2. Multiple roots (sometimes fused).

        3. The largest occlusal surface of any type of tooth.

        4. The largest size of all teeth.

Step 3—Is it upper or lower?

    An upper molar has:

        1. Three roots (may be fused).

        2. Roots arranged as follows: one lingual, one mesio-buccal, and one disto-buccal.

        3. A crown that is more square.

        4. Usually 4 or 3 cusps.

A lower molar has:

    1. Two roots (may be fused).

    2. Roots arranged as follows: one mesial and one distal.

    3. A crown that is more oblong (longer mesio-distally).

    4. Usually 5 or 4 cusps.

Step 4—What position does it hold?

    Upper Molars

      First

      1. Lingual root is largest and often widely divergent.

      2. Contact facets are found mesially and distally.

      3. Carabelli's cusp is often present.

      Second

      1. Lingual root is largest but not widely divergent.

      2. Contact facets are located mesially and distally. No distal if no M3.

      3. Carabelli's cusp is sometimes present.

      Third

      1. Roots are often fused and smaller than in 1st and 2nd.

      2. Contact facets are on the mesial surface only.

      3. Carabelli's cusp is not present.

A Carabelli's cusp is a small tubercle that is sometimes present on the mesio-lingual surfaces of the upper molars, especially the first and sometimes on the second.

    Lower Molars

      First

      1. Two separate roots, mesial curved backwards.

      2. Usually 5 cusps.

      Second

      1. Two roots may be fused, both curved backwards.

      2. Usually 4 cusps.

      Third

      1. Two fused roots that curve backwards.

      2. Variable.

Step 5—Is it right or left?

    Upper Molars (Fig. 146a)

      1. The disto-lingual cusp is the smallest.

      2. The crown is more convex on the lingual surface and when held by root with crown pointing down with *distal* surface toward you the convex side of the crown is on the side the tooth is from.

    Lower Molars (Fig. 146b)

      1. The roots are inclined toward the back.

      2. The crown is more convex on the buccal surface, and when held by roots with *distal* surface toward you, the convex side of the crown will be on the same side the tooth is from.

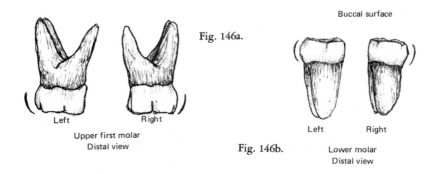

Fig. 146a.

Buccal surface

Left   Right

Upper first molar
Distal view

Left   Right

Fig. 146b.

Lower molar
Distal view

MEASUREMENTS OF THE TEETH

There are many methods of measuring teeth mentioned in the literature. Four of the most common indicators of tooth size are tooth height, mesio-distal diameter, bucco-lingual diameter, and crown module. Tooth height is frequently omitted in studies of archaeological specimens since so many prehistoric agricultural people rapidly wore down their teeth. When it is taken, it is usually expressed as the distance from the crown-root juncture to the maximum height on the crown.

The mesio-distal diameter (also called breadth, width or length in the literature) is expressed as the maximum diameter between the mesial and distal contact points.

The bucco-lingual diameter is usually expressed as the maximum diameter taken at a right angle to the mesio-distal axis.

Crown module is an expression of the relative crown mass. It is computed by averaging the mesio-distal and bucco-lingual diameters. Tooth height should ideally be included to present a complete description of crown mass but since the recording of tooth height is so severely limited by attrition, crown module is usually used for comparative and descriptive purposes.

In addition to measurements of the teeth, there are also a number of observable characteristics of the teeth are of interest to physical anthropologists. The recording of these observations is important to establish the existing variation of modern man and to see how the dentition of man has changed from that of his fossil ancestors. These observations fall into two general categories: variations in tooth number and position and variations in tooth morphology.

VARIATIONS IN TOOTH NUMBER AND POSITION

Supernumerary teeth:

Supernumerary teeth are extra teeth that may occur in the incisor, canine, premolar or molar tooth groups. Supernumerary teeth may be exact replicas of the normal teeth or they may present a morphology that does not resemble any particular tooth group. They are found in both the adult and deciduous dentitions, but only rarely in the deciduous. Frequently supernumerary teeth will occur bilaterally and may result from the retention of deciduous teeth in the adult dentition. Extra teeth may occur in many positions in or around the dental arcade.

Congenital absence of teeth:

Occasionally, one or more teeth normally present in the dentition will be missing. Of all the teeth, the third molars are most frequently missing but any of the other teeth may be congenitally absent. The investigator must be careful not to confuse congenital absence of teeth with unerupted teeth or teeth that were lost before death. Unerupted teeth may easily be detected from a radiograph. Teeth that were originally present but lost before examination present more of a problem. Of course if the teeth were lost postmortem, the cavities for the roots will be present in the alveoli. However, if a tooth was lost antemortem, the alveolus may have resorbed leaving little or no indication of the root cavity. Antemortem tooth loss can usually be detected by the characteristic distorted and unequal appearance of the alveolus. Also, if the missing tooth was at one time in contact with the adjacent teeth then wear facets may be found on the appropriate surfaces of these teeth.

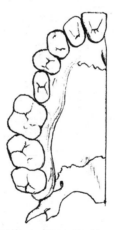

Fig. 147.   Rotated Right Maxillary
                   Second Premolar

Rotation of teeth:

Occasionally teeth will be in their proper position in the tooth row but they will appear to be rotated as much as 180° (Fig. 147). When a tooth has rotated, the normal distal surface may become the lingual surface or vice versa. The second premolars are most often rotated but other teeth can also be in this position.

TABLE 37. PERCENTAGE DISTRIBUTION OF MISSING THIRD MOLARS
AFTER DAHLBERG'S TABLE 34 (1951:171)

Group	Author		Number of Jaws	Percent Lacking One or more M_3
Chinese	Hellman	'28	19	32.0
Chinese	Knap*		64	31.2
Mongols (Buriat)	Hellman	'28	21	17.0
Japanese	Hamano*		1300	18.4
Eskimo	Hellman	28	55	13.0
Angmagssalik Eskimo	Pedersen	'49	257	26.7
East Greenland Eskimo Skulls	Pedersen	'49	81	23.5
Modern unmixed S. W. Greenland Eskimo	Pedersen	'49	210	18.6
Modern mixed S. W. Greenland Eskimo	Pedersen and Hinsch	'40	319	11.0
Labrador Eskimo	Dahlberg		23	16.0
Northwest Eskimo	Goldstein	'32	232	15.5
American Indian	Hellman	'28	55	13.0
Blackfoot Indian	Dahlberg		25	8.0
Sioux Indian	Smith*		10	50.0
Early Texas Indian	Goldstein	'48	173	19.5
Carabis Indian	Maurel*		68	70.6
Hawaiian	Dahlberg	'45	25	24.0
Melanesian	Dahlberg	'45	165	4.0
Australian Aboriginal	Hellman	'28	20	13.0
West African Negro	Hellman	'40	?	2.6
American Negro	Hellman	'28	119	11.0
European White	Hellman	'28	61	20.0
White (Hungary)	Hellman	'28	?	49.0
American White	Banks	'34	461	19.7

*Cited by Pedersen, '49.

Crowding of teeth:

Frequently the adult dentition will appear to be crowded, with one or more teeth pushed out of their normal position. This condition is usually a consequence of a reduction in size of the mandible without a corresponding reduction in the size of the teeth. The space in the alveolus is not large enough to permit the teeth to erupt in their normal positions so consequently they must erupt in altered positions. Crowding frequently accompanies impacted third molars and the rotated condition of teeth described above. Deciduous teeth are seldom crowded since they normally have plenty of room for eruption. Of all the teeth, the incisors are usually the most affected by crowding.

VARIATIONS IN TOOTH MORPHOLOGY

Molar Cusp Pattern:

The pattern of cusps and grooves on the occlusal surfaces of the molars have long been of interest to physical anthropologists in establishing differences among modern populations and in revealing our primate ancestry. This observation is severely limited when studying many prehistoric populations since the gritty nature of their diet frequently wears down their teeth at an early age and thus obscures their cusp patterns. On teeth that do display a clear cusp pattern it should be first noted that the maxillary and mandibular molars display different cusp patterns. The maxillary molars usually have 3 or 4 cusps separated by distinct grooves. In the first maxillary, the four cusps are usually about equal in size. In the second maxillary molars the fourth cusp or hypocone is usually reduced in size. In third maxillary molars the hypocone may be completely absent or be reduced to a mere cuspule on the distal surface. To record the size of these cusps Dahlberg (1951:165) has suggested the following system: "4 designates four well developed cusps; 4- indicates a reduction in the size of the hypocone; 3+ indicates absence of the hypocone, but the presence of a cuspule on the distal border; 3 indicates total absence of the hypocone" (Fig. 148).

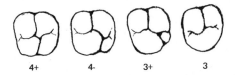

4+   4-   3+   3

Fig. 148.   Maxillary Molar Cusp Patterns

The mandibular molars normally display either four or five cusps (Fig. 149). These cusps are arranged so that the grooves between them form either a "T" or a "Y." Consequently the four types of cusp patterns are Y5, Y4, +5 and +4. Of these patterns the Y5 is common on the teeth of many fossil men, while the remaining three are recent developments of modern groups. It has been suggested that the general trend has been for the Y5 pattern to evolve into the +4 pattern through either the +5 or the Y4 stages. On modern molars the genetically stable mandibular first molar most frequently displays the ancestral Y5 pattern. Second and third molars usually display a higher frequency of the other three patterns. It should be noted that frequently cusp patterns appear intermediate in pattern and are therefore difficult to classify. Also the genetically unstable third molar frequently presents an irregular cusp pattern that does not resemble any particular type (Tables 38, 39, 40).

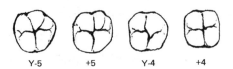

Fig. 149.

Mandibular Molar Cusp Patterns

Extra cusps:

Extra cusps have been described on many surfaces of both the maxillary and mandibular molars. Of particular interest are the extra cusps that have been termed protostylids and Carabelli's cusps (Fig. 150). A protostylid is an extra cusp that occurs on the anterior aspect of the buccal surface of the mandibular molars (Fig. 150b). According to Dahlberg (1950:24) the cusp is significant because it occurs in the South African Australopithecines, the *Meganthropus* material from Java, and the Chinese *Sinanthropus* material. He notes that is has only rarely been recorded in modern populations except in one population of 80 Pima Indians in which 37 displayed the cusp. The protostylid occurs in various degrees of prominence but it is usually recorded as being either present or absent.

Carabelli's cusp (also known as the tuberculum Carabelli and the tuberculum anomale) consists of a tubercle located on the anterior portion of the lingual surfaces of maxillary molars. The extra cusp occurs in various forms that range from a pit (Carabelli's pit) considered by some to be related to the cusp, to the well developed cusp itself. Because the cusp does occur in various forms at this

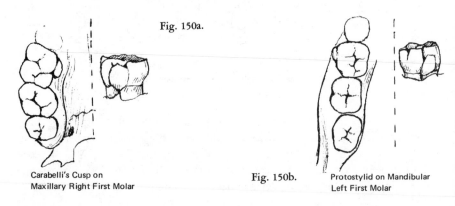

Fig. 150a.

Carabelli's Cusp on
Maxillary Right First Molar

Fig. 150b.

Protostylid on Mandibular
Left First Molar

TABLE 38. PERCENTAGE DISTRIBUTION OF PATTERNS ON LOWER FIRST MOLARS
AFTER DAHLBERG'S TABLE 25 (1951:155)

Group	Author		Number of Individuals	Y5	+5	Y4	+4
Chinese	Hellman	'28	26	100			
Mongol	Hellman	'28	36	100			
Alaska Eskimo	Goldstein	'48	67	89.6	6.0	1.5	3.0
E. G. Eskimo	Pedersen	'49	143	95.7	2.8	0	1.4
Texas Indian	Goldstein	'48	160	68.7	30.6	.6	0.0
Pecos Indian	Nelson	'37	332	88.6	10.8	0	0.6
Pima Indian	Dahlberg		162	99.4	0.6		
Ancient European White	Hellman	'28	54	83.0	0.0	11.0	6.0
European White Male	Hellman	'28	98	87.0	2.0	7.0	4.0
Chicago White	Dahlberg		75	84.0	2.0	8.0	2.0
Australian	Hellman	'28	18	100.0			
African Negro	Hellman	'28	97	99.0		1.0	

TABLE 39. PERCENTAGE DISTRIBUTION OF PATTERNS ON LOWER SECOND MOLARS
AFTER DAHLBERG'S TABLE 26 (1951:156)

Group	Author		Number of Individuals	Y5	+5	Y4	+4
Chinese	Hellman	'28	21		19.0		81.0
Mongol	Hellman	'28	39		31	5	64.0
Alaska Eskimo	Goldstein	'48	132	12.8	63.8	3.0	20.5
E. G. Eskimo	Pedersen	'49	115	19.0	42.0	4.0	35.0
Texas Indian	Goldstein	'48	206	1.5	26.2	3.4	68.9
Pecos Indian	Nelson	'37	313	8.3	24.3	1.3	66.1
Pima Indian	Dahlberg		89	2.0	69.0	1.0	28.0
Ancient European White	Hellman	'28	54	2.0	11.0	9.0	77.0
European White	Hellman	'28	110		1.0	5.0	94.0
Australian	Hellman	'28	21	5.0	43.0		52.0
African Negro	Hellman	'28	96	17	8.0	12.0	63.0

*The percentage figures of a few of the groups do not add up to 100 because the authors in some instances included other categories of patterns. However the figures are not too far off to be of comparative value.

TABLE 40. PERCENTAGE DISTRIBUTION OF PATTERNS ON LOWER THIRD MOLARS AFTER DAHLBERG'S TABLE 27 (1951:157)

Group	Worker		Number of Individuals	Y5	+5	Y4	+4	Irrg.
Chinese	Hellman	'28	16		50.0		50.0	
Mongol	Hellman	'28	31		77.0		23.0	
Alaska Eskimo	Goldstein	'48	59	20.4	69.5	0.0	10.2	
E. G. Eskimo	Pedersen	'49	55	15.0	74.0	11.0	0.0	
Texas Indian	Goldstein	'48	91	12.1	47.3	6.6	34.1	
Pecos Indian	Nelson	'37	249	8.4	51.0	4.8	35.8	
Pima Indian	Dahlberg		7		57.0		14.0	28
Ancient European White	Hellman	'28	35	6.0	34.0	11.0	49.0	
European White	Hellman	'28	74	4.0	34.0		62.0	
Australian, Aborgine	Hellman	'28	23	14.0	72.0		14.0	
African Negro	Hellman	'28	88	20.0	59.0	3.0	17.0	

location there have been several descriptive classification systems devised (Fig. 150a, Table 41). These classifications are discussed by Kraus (1959:117-123). Carabelli's cusp appears to be a recent evolutionary development that occurs in varying frequencies in all modern populations. The related Carabelli's pit has been observed in fossils dating back to the Australopithecines, but no undisputed Carabelli's cusps have been found in fossil hominoids.

TABLE 41. PERCENTAGE DISTRIBUTION OF CARABELLI'S CUSP
AFTER DAHLBERG'S TABLE 33 (1951:170)

Group	Worker	M_3 N	M_3 %	M_2 N	M_2 %	M_1 N	M_1 %
Northwest Eskimo	Dahlberg	12	0.0	17	0.0	26	7.0
Labrador Eskimo	Dahlberg	12	0.0	22	0.0	23	0.0
E. G. Greenland Eskimo Skulls	Pedersen '49	34	0.0	44	0.0	25	0.0
Living E. G. Eskimo	Pedersen '49	98	0.0	162	0.6	106	0.0
West Greenland Eskimo	Pedersen '49			None pronounced			
Central Eskimo	Pedersen '49			None pronounced			
Eskimo and White	Pedersen '49						29.4
Blackfoot Indian	Dahlberg			50	0.0	41	12.0
Pecos Pueblos	Nelson '37						8.8
Pima Indians	Dahlberg (cusps)			182	0.0	322	8.0
Pima Indians	Dahlberg (pits)			182	4.0	322	27.0
Indian Knoll	Dahlberg small (cusps)	33	6.0	33	9.0	33	24.0
American (Army)	Dietz '44					732	72.3
American Whites	Dahlberg			13	8.0	91	41.0
Swiss	M de Terra '05		1.35		.22		11.2
Dutch	Bolk '15				21.7		17.4
Lapps	Kajava '12		0.0		0.0		3.4
Bantu	Shaw '31					389	2.0

Shovel-shaped teeth:

One of the most frequently discussed genetic characters of the teeth is the shovel-shaping of the incisors (Fig. 139, Tables 42, 43). First mentioned by Hrdlicka (1907:55), its relatively high frequency of occurrence in Mongoloid racial groups has been recorded by many investigators. Morphologically, shovel-shaping involves a lingual extension of the lateral borders of the incisors. Ideally, each incisor should be classified as to the degree of shovel-shaping it exhibits on its lingual surface and the depth of the lingual fossa should be measured. However, due to the high degree of attrition on the incisors of many populations, it is often necessary to limit the classification to an observance of the presence or absence of shovel-shaping.

Occasionally an incisor will display a buccal extension of its lateral borders in addition to the usual lingual extension. The incisor displaying this phenomenon is termed a double shovel-shaped incisor. Double shovel-shaped incisors are usually maxillary but have been found in the mandible (Fig. 139).

Some incisors display such a pronounced lingual extension of the lateral borders that the tooth assumes the appearance of a distorted premolar or a barrel shape. Barrel shaping most often occurs on maxillary lateral incisors.

Peg-shaping:

Occasionally teeth will occur that appear abnormally small and unlike their normal appearance (Fig. 151). Such teeth have been described as peg-shaped and occur most frequently on the genetically unstable third molars and lateral incisors. Peg-shaping has been reported in all modern populations and appears to be related to congenital absence.

Peg-shaped Maxillary Right
Third Molar

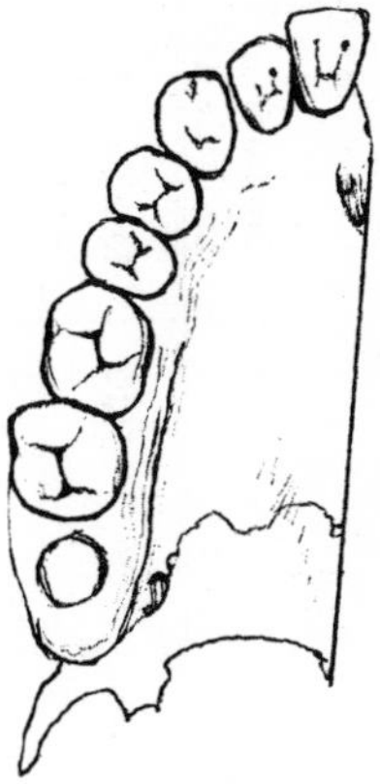

Fig. 151.

Taurodontism:

Taurodontism is a condition found in the molars in which the pulp cavity is enlarged and the roots are reduced (Fig. 152a). This condition has frequently been observed in Neanderthal and other fossil forms. In addition it has been found to occur occasionally in some modern populations. Taurodontism occurs in various degrees of prominence which have been classified by Middleton Shaw (1928) into four types: cynodont, hypotaurodont, mesotaurodont, and hypertaurodont.

TABLE 42. PERCENTAGE FREQUENCY OF SHOVEL-SHAPED INCISORS
AFTER DAHLBERG'S TABLE 22 (1951:144)

Median Incisors

Group	Author		Number of Individuals	Shovel-shaped marked	semi	Total of marked & semi-shoveled	Trace	None
Chinese	Hrdlička	'20	female 208 male 1094	82.7 66.2	12.5 23.4	94.2 89.6	1.0 1.8	3.8 7.8
Mongolian	Hrdlička	'20	24	62.5	29.0	91.5	8.5	
Eskimo	Hrdlička	'20	40	37.5	47.5	84.0	15.0	
E. G. Eskimo	Pedersen	'49	116	83.6	14.7	95.3		
Mixed Indians	Wissler	'31	male 1388 female 1205			85.0 85.0		
Pima Indians	Dahlberg		female 125 male 101	99.0 96.0		99.0 96.0	1.0 4.0	
Pueblos Indians	Dahlberg		21	81.0	19.0	100.0	0.0	
Sioux	Hrdlička	'31	116	98.3		98.3	1.7	
Pecos Pueblos	Nelson	'37	324	74.1	15.4	89.5	8.3	2.2
Pecos Pueblos	Hooten	'30	124	86.3		86.3	13.7	
Early Texas Indian	Goldstein	'48	124	95.1		95.1	4.9	
Indian Knoll	Dahlberg & Snow		30	84.0	16.0	100.0		
American Negro	Hrdlička	'20	male 618 female 1000	4.9 3.6	7.6 8.0	12.5 11.6	33.0 32.6	54.5 56.0
American White	Hrdlička	'20	male 1000 female 1000	1.4 2.6	7.6 5.2	9.0 7.8	24.5 21.8	66.5 70.4

TABLE 43. PERCENTAGE FREQUENCY OF SHOVEL-SHAPED INCISORS
AFTER DAHLBERG'S TABLE 23 (1951:145)

Lateral Incisors

Group	Author		number of Individuals	Shovel-shaped marked	semi	Total of marked & Semi-shoveled	Trace	None	Anomalous form
Chinese	Hrdlička	'20	female 208 / male 1094	68.8 / 56.9	13.5 / 24.0	82.3 / 90.9	1.0 / 1.5	3.4 / 9.5	
Mongolian	Hrdlička	'20	24	75.0	25.0	100.0			
Eskimo	Hrdlička	'20	37	57.0	43.0	100.0			
Indians	Hrdlička	'20	300	76.0	17.0	93.0	6.0	1.0	
Indians	Wissler	'31	Male 1356 / female 1186	82.0 / 87.0		82.0 / 87.0			
Pecos Pueblos	Nelson	'37	338	72.0	17.4	89.4	9.3	1.4	
Pueblos Indians	Dahlberg		21	81.0	14.0	95.0			5.0
Pima	Dahlberg		female 119 / male 93	81.0 / 81.0		81.0 / 81.0	7.0 / 13.0	1.0	12.0 / 5.0
Indian Knoll	Dahlberg & Snow		30	80.0	17.0	97.0			3.0
American Negro	Hrdlička	'20	male 618 / female 1000	4.5 / 3.8	12.8 / 11.1	17.3 / 14.9	38.0 / 35.1	42.1 / 47.5	
American White	Hrdlička	'20	male 1000 / female 1000	1.4 / 1.0	8.8 / 7.4	10.2 / 8.4	36.4 / 29.9	50.0 / 59.6	

Fig. 152a.

Normal   Taurodont

Enamel extensions and pearls:

In some molars and premolars an extension of the crown enamel will occur between the roots (Fig. 152b). Sometimes this extension will culminate in a cluster of enamel termed an enamel pearl. If the teeth are imbedded in the alveolus, the enamel pearls are often obscured from observation by the surrounding bone. Enamel pearls have been observed in many human populations but little is known of their frequency of distribution or of their evolutionary significance.

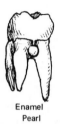 

Fig. 152b.   Enamel Pearl   Enamel Extension

Extra or missing roots:

Occasionally a tooth will display an extra root or the absence of a root that is normally present. Is is important to record the number of roots present whenever possible.

Moving from the first molar to the third, one can see a tendency for the molar roots to become less divergent and more fused. A similar trend occurs in the premolars.

Occlusal wear:

During the process of mastication, the mandibular and maxillary teeth continually rub against each other and against whatever gritty food particles are in the mouth cavity. This continual abrasive action eventually wears down the occlusal surfaces of the teeth, destroys the cusp patterns on molar crowns and eventually exposes the underlying dentin. The rate at which this destructive process of attrition takes place greatly depends on the diet of the population involved. In most aboriginal American Indian populations for example the process of attrition proceeds quite rapidly. It has been suggested that this is due to the excessive amount of grit in their diet that probably originated from their grinding stones. In contrast, most modern Americans experience a very retarded rate of attrition; again because of their diet. The degree of attrition can provide effective criteria for determining the age at death of an individual as long as the rate of attrition of that particular population is known. Many investigators rely principally on the degree of attrition on the molars to determine age. The molars are particularly useful because the cusp patterns on their occlusal surfaces allow several progressive stages of attrition to be identified and correlated with age at death. When the molars are used, it must be remembered that attrition will not be marked to the same degree on all molars since they erupt at different ages. In other words the first molars are exposed to about twelve more years of mastication than the third molars and about six years more than the second molars. When an age determination is attempted that difference needs to be kept in mind. Again, it is important to remember that all populations do not have the same rate of attrition and therefore the criteria of determining age from tooth wear in one population does not necessarily apply to another population. Unfortunately, all of the dentitions within a population do not wear at the same rate due to individual differences in diet and tooth structure. This severely limits the accuracy of age determination by this method and other criteria should be consulted whenever possible.

The following is D. R. Brothwell's age classification of wear on pre-medieval British teeth:

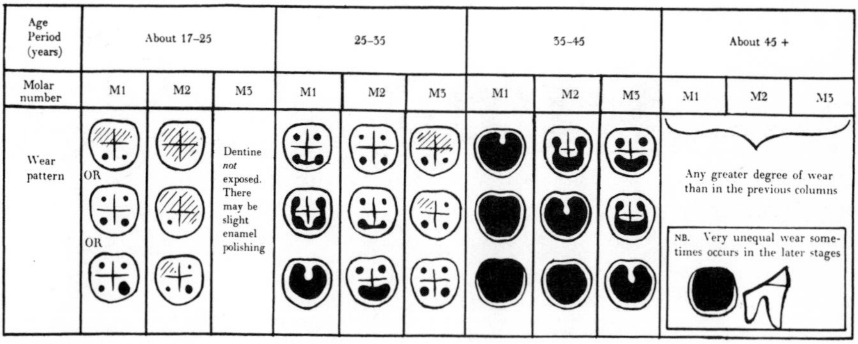

Fig. 153. A correlation of age at death with molar wear in pre-medieval British skulls (After Brothwell, 1965:69) Permission for reproduction granted by Dr. Donald R. Brothwell and the Trustees of the British Museum of Natural History.

Occlusion:

The type of occlusion that exists between the maxillary and mandibular teeth has been of interest to anthropologists (Fig. 154). In adult skulls, Mongoloids most frequently exhibit edge-to-edge occlusion of the teeth. In contrast Caucasoids most frequently display a slight overbite so that the maxillary incisors project more anteriorly than those of the mandible. Underbites are uncommon in all racial groups but can occasionally be found.

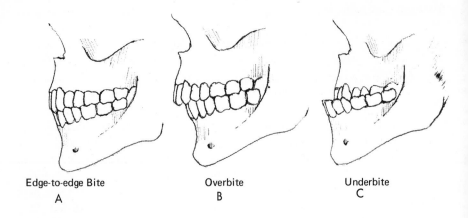

Edge-to-edge Bite
A

Overbite
B

Underbite
C

Fig. 154. Occlusion of maxillary and mandibular teeth

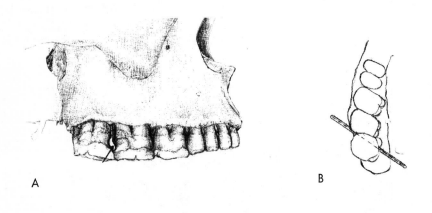

A

B

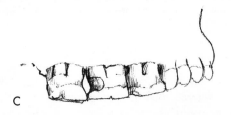

C

Fig. 155. Artificial deformation by grooves between the teeth.

Artificial deformation:

Several features of tooth morphology may be attributed to cultural sources. In some cases the teeth have been intentionally filed or chipped to produce an ornamental effect. Dental mutilations in America have been classified into seven types by J. Romero (1958). These were principally found in Mexico. Several earlier studies discuss the world distribution of dental mutilation, reasons for its practice, and techniques involved. They describe mutilations from Africa, Australia, Malaya and Egypt in addition to the examples from the New World. The incisors are the teeth usually involved since they are the teeth most visible from outside the mouth.

Ubelaker et al. (1969) have reported the occurrence of artificial grooves located interproximally in the molar regions of several American Indian populations. These grooves usually occur near the junction of the crown and root of the tooth involved (Fig. 155). The frequent association of these grooves with carious lesions, alveolar abscesses, and alveolar resorption resulting from periodontal disease suggested that these grooves were produced in an attempt to relieve discomfort in the immediate area. The investigators observed the phenomenon in material from archaeological sites geographically distributed over much of the United States and temporally distributed from archaic to recent. This type of deformation is easily overlooked and care should be taken to describe its occurrence.

APPENDIX I

CLASSIFICATION OF BONES ACCORDING TO SHAPE

LONG BONES—such as the femur, humerus and radius consist of a shaft (or diaphysis) and two extremities (or epiphyses).

SHORT BONES—such as those of the metacarpus (hand) and metatarsus (foot) consist of a shaft and two extremities and are sometimes referred to as long bones in miniature.

FLAT BONES—such as those of the skull, scapulae, and innominates and have relatively extensive surfaces for protection or for muscular attachment.

IRREGULAR BONES—such as the vertebral, maxillary and sphenoid bones. These have irregular, peculiar and often very complex shapes.

PARTS OF BONES

ALA—a winglike structure

APOPHYSIS—a prominence formed directly upon a bone

CANAL—narrow passage or channel

CAPITULUM—a small rounded articular end of a bone

CONDYLE—a rounded projection on a bone usually for articulation with another bone

CONOID—cone shape

CORONOID—shaped like a crow's beak

CREST—a sharp border or ridge

DENS—a tooth

DIAPHYSIS—the shaft of a bone

DIPLOË—latticelike bone, *e.g.* tissue between the tables of the skull

EPICONDYLE—above a condyle

EPIPHYSIS—the extremity of a bone expanded for articulation

FACET—a small, smooth area on a bone

FORAMEN—a hole

FOSSA—a depression

FOVEA—a pit or cuplike depression

GLENOID—having the appearance of a socket

HEAD—a rounded, smooth eminence for articulation

INCISURE—a notch

LIP—margin of a groove, crest or line

METAPHYSIS—a line of junction between epiphysis and diaphysis

PIT—a tiny hole or pocket

PROCESS—any kind of projection

PROMONTORY—a projecting part

RIDGE—a rough, narrow elevation extending some distance

SINUS—a cavity in bone lined with mucous membrane

SPINE—a sharp prominence

STRIA—a line

STYLOID—resembling a stylus

SULCUS—a groove

SUTURE—a form of articulation found only in the skull

TROCHANTER—a large prominence for attachment of rotator muscles

TROCHLEA—a pulley

TUBERCLE—a small tuberosity

TUBEROSITY—a rounded eminence

DEFINITIONS OF OSTEOLOGICAL TERMS

ACETABULUM—cavity on external surface of hip bone for head of femur

ACROMION—tip of shoulder

ASTRAGALUS—ankle bone

ATLAS—first cervical vertebra

AURICULAR—sacro-iliac articulation

AXIS—second cervical vertebra

BASILAR—suture on base of skull between occipital and sphenoid

BOSS—a rounded eminence

CALCANEUS—heel bone

CAPITATE—head

CARPUS—wrist

CERVICAL—neck

CHONDRAL—pertaining to cartilage

CLAVICLE—key, collar bone

COCCYX—tail bone

CONCHA—shell

CORACOID—like a crow's beak

CORONAL—suture between frontal and parietals

CORPUS—body

COSTA—rib, side

COXA—hip

CUBOID—cube-like

CUNEIFORM—wedge-shaped

DIGIT—finger or thumb

ETHMOID—sieve-like

EPISTROPHEUS—second cervical vertebra

FEMUR—thigh bone

FIBULA—brace bone

HAMATE—hooked

HUMERUS—upper arm bone

HYOID—U-shaped bone

ILIUM—hip, haunch

INCUS—anvil

INNOMINATE—unnamed

INTEROSSEOUS—occurring between bones

ISCHIUM—hipbone

LACRIMAL—tear

LAMBDOIDAL—suture between parietals and occipital

LINEA ASPERA—longitudinal ridge on the posterior surface of the femur

LUMBAR—loin

LUNAR—moon-shaped

MALAR—cheek

MALLEOLUS—little hammer

MALLEUS—hammer

MANDIBLE—lower jaw

MANUBRIUM—handle

MASTOID—breast-like

MAXILLA—upper jawbone

MEATUS—an opening

MENTAL FORAMEN—on body of mandible

METACARPAL—beyond the wrist

METATARSAL—beyond instep

MULTANGULAR—many angles

NASAL—nose

NAVICULAR—boat-shaped

NUTRIENT FORAMEN—hole in a bone through which nutrients are received

OBTURATOR—the major foramen of the hip bone

OCCIPITAL—base of head

OCCLUSAL PLANE—masticating surfaces of the teeth

OLECRANON—process of ulna at elbow

ORBIT—bony socket which contains the eye

PALATE—roof of mouth

PARIETAL—wall

PATELLA—knee pan or pan

PELVIS—basin

PHALANGES—line of soldiers

PISIFORM—pea-shaped

POPLITEAL—posterior surface of the knee

PUBIS—pubic bone

RADIUS—spoke or ray
RESORPTION—removal by absorption
SACRUM—holy bone
SAGITTAL—shaped like arrow, straight
SCAPHOID—boat-shaped
SCIATIC—a notch in the hip bone
SEMILUNAR—half moon
SESAMOID—an oval nodule of bone
SPHENOID—wedge-shaped
SQUAMOSAL—platelike
STAPES—stirrup
STERNUM—flat, breast bone
STYLOID—long and pointed

SUSTENTACULUM—a support
SYMPHYSIS—a point of junction
TALUS—ankle bone
TARSUS—instep
TEMPORAL—time, instep
THORAX—chest, cage
TIBIA—shin bone, flute
TRAPEZIUM—a little table
TRAPEZOID—table-like
TRIQUETRAL—triangular
ULNA—elbow
VERTEBRA—to turn, spindle
VOMER—ploughshare
XIPHOID—like a sword
ZYGOMATIC—cheek

TERMS INDICATING THE SIDE AND DIRECTION OF THE PARTS OF THE BODY

ANTERIOR—in front
AXILLARY—toward the armpit
CAUDAL—toward tail
CRANIAL—toward head
DISTAL—farthest from center
DORSAL—back
EXTERNAL—outside of
FRONTAL—in front
INFERIOR—lower
INTERNAL—inside of
LATERAL—to side, away from midline
MEDIAL—toward the midline
PLANTAR—sole of the foot

POSTERIOR—behind
PROFUNDUS—deep
PROXIMAL—nearest the center
RADIAL—lateral view of metacarpals
SUPERFICIAL—near surface
SUPERIOR—above
TRANSVERSE—crosswise
ULNAR—medial view of metacarpals
VENTRAL—in front
VERTEX—top
VOLAR—relating to the palm or sole of the foot

APPENDIX 2

BONES AND THEIR PLURALS

BONES AND THEIR PLURALS

Name of Bone	Plural
Cranium	
ethmoid	ethmoids
frontal	frontals
occipital	occipitals
palate	palate bones
parietal	parietals
sphenoid	sphenoids
temporal	temporals
Face	
concha	conchae
lacrimal	lacrimals
mandible	mandibles
maxilla	maxillas or maxillae
nasal	nasals
vomer	vomers
zygomatic (malar)	zygomatics (malars)
Ear	
hyoid	hyoids
incus	incudes
malleus	mallei
stapes	stapeses
Vertebral Column	
coccyx	coccyges
sacrum	sacra
vertebra	vertebrae
Thorax	
gladiolus	gladioluses or gladioli
manubrium	manubriums or manubria
rib	ribs
sternum	sterna or sternums
xiphoid	xiphoids

Upper Extremities

capitate	capitates
carpal	carpals
clavicle	clavicles
greater and lesser multangular	multangulars
hamate	hamates
humerus	humeri
lunate	lunates
navicular	naviculars
phalanx	phalanges
pisiform	pisiforms
scapula	scapulae
triquetrum	triquetrums
ulna	ulnae

Lower Extremities

calcaneus	calcaneuses or calcanea
cuboid	cuboids
cuneiform	cuneiforms
femur	femora
fibula	fibulas or fibulae
ilium	ilia
innominate	innominates
ischium	ischia
patella	patellae
pubis	pubes
symphysis	symphyses
talus	tali
tarsal	tarsals
tibia	tibias or tibiae

Other Terminology

foramen	foramina
epiphysis	epiphyses
process	processes
ala	alae
pelvis	pelves
stria	striae
alveolus	alveoli

WORD ANALYSIS FOR STUDENTS IN SCIENCE

A. Prefixes

A prefix is a short word form used at the beginning of a word which modifies the meaning of the word. Often the spelling of a prefix is changed to make pronunciation easier. For example, the Latin prefix ad is changed so that instead of adpendage we have appendage. Some of the most common prefixes used in biological terms, which are derived from the Old English, Greek and Latin, are listed below with some variations in spellings, meanings and examples.

Prefix	Meaning	Examples
	(From Old English)	
fore-	before, in front	forearm
un-	not	unconscious
	(From Greek)	
a-, an-	without, lacking	asexual
		anaerobic
amphi-	on both sides	amphibian
ana-	up	anatomy
anti-	against	antiserum
cata-	down	catabolism
dia-	through	diaphragm
epi-	over	epidermis
hyper-	excessive	hyperthyroidism
hypo-	under	hypothalamus
meta-	after, change	metaphase,
		metamorphic
para-	beside	parabasal
peri-	around	perianth
pro-	for, before, in front of	pronotum
syn-, sym-, sys-	together	synapsis,
		sympetalous,
		systole
	(From Latin)	
ab-, abs-	from, away	abduct, abscess
ad-, af-, ag-	toward, to	adduct, afferent,
		agglomerate
bi-	two	bifocal
circum-	around	circumflex
com-, con-	with	commensal,
		conjugation

Prefix	Meaning	Examples
de-	down, away from, separation	depressed
dis-, dif-	away, apart	disarticulation, diffusion
ex-, ef-	out, from	extraction, efferent
extra-	outside	extracellular
in-, im-	in, within	inclusion, immersion
in-	not	incapacitate
inter-	between	internode
intra-	within	intracellular
ob-, oc-	over, toward	obtected, occlusion
post-	after	postscutellum
pre-	before	prenatal
pro-	for, before, in front of	prophase
re-	back, again	regression, refracture
semi-	half	semicircular
sub-, sus-	under	subcutaneous, suspend
super-, supra-	over, extra, above	supersensitive, suprarenal
trans-	across	transpiration
ultra-	beyond	ultrasonic

B. ROOTS

A root is a word form that can not be analyzed and has a relatively constant form and meaning. It is sometimes difficult to distinguish between prefixes and roots, but a root is usually more important and may be found at various positions in words. If you know the meaning of the root or roots which are parts of a word you may often obtain a general idea of the meaning of that word. Most of the roots used in biological terms are derived from Latin or Greek. Scientists have made use of both Greek and Latin roots to form combinations to produce new words for previously unknown phenomena. There are often meaningless, short connectives between word parts such as in the word chromosome which can be analyzed as—chrom-, color, -o-, the meaningless connective and -soma, body.

The following list contains some of the more common roots from Greek and Latin which are used as basis for biological terminology. The root, meaning and examples are given.

Root	Meaning	Examples

(From Greek)

Root	Meaning	Examples
-anthrop-	man	anthropomorphic
-aster-	star	Asteroidea
-auto-	self	autonomic
-bio-	life	biogenesis
-chrom-	color	chromatin
-chloro-	green	chloroplast
-cyto-	cell	cytology
-derm-	skin	dermis
-ecto-	outside	ectoplasm
-endo-	within	endodermis
-gastro-	stomach	gastrovascular
-hem-	blood	hemolysis
-hetero-	different	heterozygous
-homo-	same	homologous
-hydr-	water	hydrobiology
-leuco-	white	leucocyte
-mega-	large	megaspore
-meso-	middle	mesoglea
-micr-	small	microscope
-mon-	one	mononucleate
-morph-	form	morphology
-orth-	straight	Orthoptera
-phor-	bearing	Mastigophora
-phot-	light	photosynthesis
-plasm-	a thing moulded	endoplasm
-pod-	foot	Gastropoda
-poly-	many	polymorphism
-proto-	first	protoplasm
-pseud-	false	pseudopodium
-pter-	wing	Hymenoptera
-som-	body	somatic
-tri-	three	triploid
-zo-	animal	zoology

(From Latin)

Root	Meaning	Examples
-ac-	sharp	acute
-aqu-	water	aqueous
-aud-	hear	auditory
-brev-	short	brevicornis
-capit-	head	capitulum
-carn-	flesh	carnivore
-cid-, -cis-	kill, cut	insecticide
		excise

Root	Meaning	Examples
-corp-	body	corpuscle
-dec-	ten	Decapoda
-dent-	tooth	dentate
-duc-	lead	adduct
-flor-	flower	flora
-gen-	origin, kind	gene, genus
-loc-	place	locus
-mar-	sea	marine
-multi-	many	multiple
-mut-	change	mutation
-nomen-	name	nomenclature
-omni-	all	omnivorous
-ped-	foot	millipede
-seg-, -sect-	cut	segmental, dissection
-spir-	breathe	spiracle
-terr-	land	terrestrial
-uni-	one	unicellular
-vac-	empty	vacuum
-vol-	wish	voluntary
-volv-, -volu-	roll, turn	Volvox, evolution

APPENDIX 3

EXCAVATION AND TREATMENT OF SKELETAL REMAINS

One of the major reasons for such a text as this is to convey to the student some of the proper procedures that should always be followed when (1) excavating, (2) transporting, (3) cleaning, (4) restoring and (5) handling skeletal material. It is of equal importance to stress things that should never be done.

To some readers many of the things I will caution against may seem stupid, but they do occur. Much bone damage and mixing of skeletal material comes from thoughtless actions. Always think of the consequences of your actions both in the field and in the laboratory.

Excavation

ALWAYS:

1. Leave bones in place until whole skeleton is exposed.
2. Take photographs of exposed skeleton before removal.
3. Place markers for site and feature identification, direction (north arrow), and distance (centimeter or inch scale) alongside skeleton before photographing.
4. Keep accurate records (see attached burial sheet). I have excavated more than 2500 burials in the past 12 years and have used many types of record forms, but the most complete is that used by the River Basin Survey of the Smithsonian Institution. With their permission, this form is reproduced on the following page. Gather the specified data when the complete skeleton has been exposed but before removal of any part. Do not rely on memory.

RIVER BASIN SURVEYS
BURIAL FORM

Feature No. _____   Site No. _____

Burial No. _____   State _____

Reservoir _____   County _____

1. LOCATION

   a. Square _____

   b. Horizontal _____

   _____

   c. Depth from surface _____

   d. Depth from datum _____

2. BURIAL TYPE

   a. Extended _____   d. Reburial _____

   b. Flexed _____   e. Cremation _____

   c. Semiflexed _____   f. Part crem. _____

   g. Other _____

3. BURIAL DIMENSIONS

   a. Max. length _____   Dir. _____

   b. Max. width _____   Dir. _____

   c. Thickness _____

4. DEPOSITION

   a. Position _____

   b. Head to _____

5. GRAVE TYPE

   a. Surface _____   c. Cist _____

   b. Pit _____   d. Other _____

   e. Shape _____

6. GRAVE DIMENSIONS

   a. Max. length _____   Dir. _____

   b. Max. width _____   Dir. _____

   c. Depth _____

7. STRATIFICATION

   a. Inclusive _____   c. Precedent _____

   b. Intrusive _____   d. Disturbed _____

   e. _____

8. ASSOCIATIONS

   a. Features _____

   _____

   _____

   _____

   _____

   _____

   _____

   b. Specimens _____

   _____

   _____

   _____

   _____

   _____

   _____

9. PRESERVATION

   a. Poor _____   Fair _____   Good _____

10. COMPLETENESS

   _____

11. SEX

   a. M ____   F ____   Indeterminate ____

12. AGE

   a. Infant _____   d. Adult _____

   b. Child _____   e. Mature _____

   c. Adolescent _____   f. Senile _____

13. NEG. Nos. _____

14. REMARKS _____

   _____

   _____

Site No.

Feature No.

Burial No.

Reservoir

Recorded by _____   Date _____

GPO WFSO 8-19-48 2000 50-154

5. Use your trowel in a sweeping motion.
6. Place bones in marked containers (site number, burial number, feature number, date, excavator etc.).
7. Use several bags to sack a complete skeleton, one for the skull and mandible, one for big bones (femora, tibiae etc.) and one for small bones (vertebrae, ribs etc.). By placing the bones of each hand and each foot in separate bags their subsequent identification will be facilitated.
8. Mark containers with a waterproof ink.
   Note: Any container will do as long as it is carefully marked.

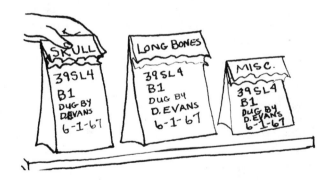

9. Brush as much dirt from the bones as possible at the time of recovery. Use a small paint brush.
10. Allow wet or damp bones to dry for a few hours in shade before removal; damp bones are easily broken. Allowing bones to dry in direct sunlight, especially in mid-summer, may cause longitudinal cracking.
11. Keep every piece of bone. They will aid in the restoration later and will increase the accuracy of the analysis.

NEVER

1. Stick a trowel into the ground to pry out bone. You cannot see what damage the end of the trowel is doing.
2. Leave dirt in the skull. It will harden, shrink and act as a cannon ball to crush the bone. Even if the skull breaks, it can be glued back together if all the pieces are kept.

3. Pull a partly exposed bone out of the ground. Wait until the skeleton is completely excavated. By removing the bone too soon you will not only lose scientific information but may break off the end. This will create more repair work later.

4. Place skeletal material in unmarked bags.
5. Treat bone with preservative of any kind in the field. Doing so will cause dirt to adhere to the bones and will necessitate extra work later in cleaning and restoration. If you wish to save bones or a skeleton *in situ,* remove by placing a plaster jacket around the bones and supporting earth. A water soluble glue like Elmer's can be used on very fragile material.

Transportation

ALWAYS

1. Pack marked bags of bones carefully. If bags are packed in a carton or box, place the open ends of the sacks at different ends of the box. If skeletal material is jarred out, it will then be easier to replace it in the proper sack.
2. Place the bones in small cartons with one or two burials to a carton when moving a large series of bones.
3. Take care to make sure bones will not shake out of bags or boxes and be lost or mixed.

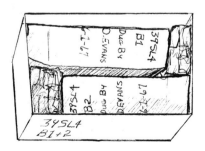

NEVER

1. Toss bones in the back of a car or truck without proper packing and drive over rough roads.
2. Pack a series of skeletons in one large container (the size of a refrigerator, stove or deep freeze).
3. Pack rocks or heavy artifacts on top of bones.

Cleaning

ALWAYS

1. Clean material thoroughly over a screen-bottomed tray using water or acetone with a brush (typewriter brush, paint brush etc.). If you use a tooth brush, do not use one with nylon or plastic bristles. Method of cleaning will depend on the condition of the bone. If the bone is solid, hard and in good condition, wash it in water with a soft brush. If the bone is moist, soft and flaky, allow it to dry under a fan, and brush dirt from the surface.

2. Be careful when washing or cleaning bones and teeth not to lose them down the drain. Always use a screen-bottomed tray. Teeth are easily lost.

3. Label every cleaned bone with site number, burial number, catalogue number, or some appropriate designation, so that material will not become mixed. Use waterproof ink.

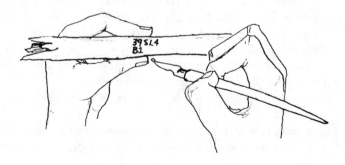

NEVER

1. Damage joints by rough brushing while attempting to remove dirt that is hard to get at.
2. Wash more than one skeleton at a time, as mixing of the material may result.

3. Sand or file off broken edges to make them join.
4. Glue a tooth in a socket until the wear surfaces are matched (if there is wear surface; loss of teeth may eliminate this valuable aid).

Restoration

ALWAYS

1. Wait until all of the bones are dry. When they are dry, broken bones can be restored by gluing them together with Duco cement or Ambroid (water soluble glues are not recommended because they absorb moisture and allow the pieces to come apart). A box or pan of fine sand is needed to support the mended bones while the glue is hardening. Modeling clay or plasticene, wooden matchsticks and light wire can be used to support the mended bones when necessary.
2. Assemble smaller parts into larger pieces, and then fit the larger pieces together.
3. Be absolutely certain that you are gluing together the right pieces. Color and texture of the pieces should be considered in matching broken edges.
4. Label all bones at the end of the cleaning process.
5. Follow cleaning and restoration of the bones with dipping them in a transparent preservative such as Alvar, Duco, Ambroid or Gelva mixtures. The mixture should be the consistency of water.

NEVER

1. Work on more than one skeleton at a time unless all bones are properly numbered. This will eliminate commingling of bones from different skeletons.
2. Fill in broken areas with plaster or plastic wood. This may cover evidences of pathology or of cultural practices such as cut marks on bones.
3. Cover bones with oil, paint or shellac, as this covers up suture patterns and possible pathological processes.
4. Glue together unclean edges.
5. Use rubber cement to glue bones together because it rapidly loses its adhesive property.

Handling

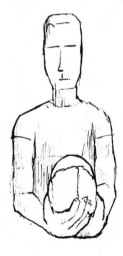

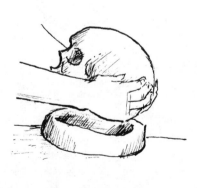

ALWAYS

1. Handle any bone carefully.
2. Handle skulls with both hands; never put fingers in eye orbits.
3. Use a bean bag or donut ring when placing the skull on a table.
4. Replace bones in their proper storage location.

NEVER

1. Pick-up skulls by eye orbits.
2. Drop bones—they break!
3. Walk into a laboratory and pick up any item—bone or artifact—unless you ask permission. If you pick up the bone from an assortment on a table, keep a finger on the spot where the bone was removed. You will then know exactly where it came from and will avoid getting it out of order.

SELECTED REFERENCES

Akabori, E.
  1934 Septal Apertures in the Humerus in Japanese, Ainu and Koreans. *American Journal of Physical Anthropology,* 18: 395-405.

Anderson, J. E.
  1962 The Human Skeleton: A Manual for Archaeologists. National Museum of Canada

Anderson, Margaret, M. B. Messner and W. T. Green
  1964 Distribution of Lengths of the Normal Femur and Tibia in Children from One to Eighteen Years of Age. *The Journal of Bone and Joint Surgery,* 46A: 1197-1202.

Bass, William M.
  1969 Recent Developments in the Identification of Human Skeletal Material. *American Journal of Physical Anthropology,* 30: 459-462.

Boucher, B.J.
  1955 Sex difference in the foetal sciatic notch. Journal of Forensic Medicine, Vol. 2, pp. 51-54.

  1957 Sex differences in the foetal pelvis. American Journal of Physical Anthropology, Vol. 15, pp. 581-600.

Breitinger, E.
  1938 Zur Berechnung der Körperhöhe aus den langen Gliedmassenknochen Anthropologischer Anzeiger, Vol. 14, pp. 249-274.

Brekhus, P. J., C. P. Oliver and G. Montelius
  1944 Study of Pattern and Combination of Congenitally Missing Teeth in Man. *Journal of Dental Research,* 23: 117.

Brothwell, D. R. (Editor)
  1963 Dental Anthropology. Vol. 5 *Symposium of the Society for the Study of Human Biology.* Pergamon Press, New York.

Brothwell, D. R.
  1965 Digging up Bones. British Museum, London.

Butler, P. M.
  1937 Studies of the Mammalian Dentition. I. The Teeth of *Centetes ecandatus* and its Allies. *Proceedings of the Zoological Society of London,* 107: 103.

  1939 Studies of the Mammalian Dentition. Differentiation of the Post-cranial Dentition. *Proceedings of the Zoological Society of London,* 109: 1-36.

  1961 Relationships Between Upper and Lower Molar Patterns. *International Colloquium on the Evolution of Mammals.* Part 1. Kon. Vlaamse Acad. Wetensch Lett. Sch. Kunsten Belgie, Brussels, pp. 117-126.

  1963 Tooth Morphology and Primate Evolution. In *Dental Anthropology.* D. R. Brothwell editor. Pergamon Press Ltd., London, pp. 1-14.

Campbell, T. D.
  1925 Dentition and Palate of the Australian Aboriginal. The Hassell Press, University of Adelaide.

Coleman, William H.

1969 Sex Differences in the Growth of the Human Bony Pelvis. *American Journal of Physical Anthropology,* 31: 125-151.

Cornwall, I. W.

1956 Bones for the Archaeologist. The Macmillan Company, New York

Dahlberg, A. A.

1945 The Changing Dentition of Man. *Journal of the American Dental Association.* 32:676-690.

1949 The Dentition of the American Indian. *The Physical Anthropology of the American Indian.* Viking Fund Inc., New York, pp 138-176.

1950 The Evolutionary Significance of the *Protostylid. American Journal of Physical Anthropology,* 8: 15-27.

1951 The Dentition of the American Indian. *The Physical Anthropology of the American Indian.* Viking Fund, Inc., 5: 138-176.

1963 Dental traits as identification tools. *Dental Progress,* 3: 155-160.

Dahlbert, A. A. and Oscar Mikkelson

1947 The Shovel-shaped Character in the Teeth of the Pima Indians. *Proceedings of the 16th Annual Meeting, American Journal of Physical Anthropology,* Vol. 5, No. 2.

Dupertuis, C.W. and John A. Hadden, Jr.

1951 On the reconstruction of stature from long bones. American Journal of Physical Anthropology, Vol. 9, pp. 15-53.

Dwight, Thomas

1894 Methods of estimating the height from parts of the skeleton. Medical Records, New York, Vol. 65, pp. 16-24.

1905 The Size of the Articular Surfaces of the Long Bones as Characteristics of Sex An Anthropological Study. *American Journal of Anatomy* 4: 19-31.

Garn, Stanley M., Arthur B. Lewis and D. L. Polacheck

1959 Variability of tooth formation. *Journal of Dental Research,* 38: 135-148.

Garn, Stanley M., C. G. Rohmann and T. Blumenthal

1966 Ossification Sequence Polymorphism and Sexual Dimorphism in Skeletal Development. *American Journal of Physical Anthropology,* 24: 101-116.

Genovés, Santiago C.

1959 Diferencias Sexuales en al Hueso Coxal. U. National Autonoma de Mexico. Public. del Inst. de Hist. Primera serie, No. 49, Mexico City, D.F.

1966 La Proporcionalid entre Los Huesos Largos y su Relacion con la Estatura en Restos Mesoamericanos. Instituto de Investigaciones Historicas. Serie Antropologica No. 19.

1967 Proportionality of long bones and their relation to stature among Mesoamericans. American Journal of Physical Anthropology, Vol. 26, pp. 67-78.

1971 Personal Communication, March 15, 1971.

Giles, Eugene

    1964 Sex Determination by Discriminant Function Analysis of the Mandible. *American Journal of Physical Anthropology,* 22: 129-135.

    1966 Statistical Techniques for Sex and Race Determination: Some Comments in Defense. *American Journal of Physical Anthropology,* 25: 85-86.

    1970a Sexing Crania by Discriminant Function Analysis: Effects of Age and Number of Variables. *In* Proceedings VIIIth International Congress of Anthropological and Ethnological Sciences, Tokyo and Kyoto. *Anthropology,* I: 59-61.

    1970b Discriminant Function Sexing of the Human Skeleton. *In* Personal Identification in Mass Disasters. Edited by T. D. Stewart. National Museum of Natural History, Smithsonian Institution, Washington, pp 99-109.

Giles, Eugene and O. Elliot

    1962 Race Identification from Cranial Measurements. *Journal of Forensic Sciences,* 7: 147-157.

    1963 Sex Determination by Discriminant Function Analysis of Crania. *American Journal of Physical Anthropology,* 21: 53-68.

Goaz, Paul W. and M. Clinton Miller III

    1966 A Preliminary Description of the Dental Morphology of the Peruvian Indian. *Journal of Dental Research,* 45: 106-119.

Goldstein, Marcus J.

    1948 Dentition of Indian Crania from Texas. *American Journal of Physical Anthropology,* 6: 63-84.

Grant, I. C. B.

    1952 A Method of Anatomy. The Williams and Wilkins Co., Baltimore.

Graves, William W.

    1921 The Types of Scapula. *American Journal of Physical Anthropology,* 2: 111-128.

    1922 Observations on Age Changes in the Scapula: A Preliminary Report. *American Journal of Physical Anthropology,* 5: 21-31.

Gregory, W. K.

    1926 Palaeontology of the Human Dentition: 10 Structural Stages in the Evolution of the Cheek Teeth. *American Journal of Physical Anthropology,* 9: 401-426.

    1926 The Origin and Evolution of the Human Dentition. Williams and Wilkins Co., Baltimore.

Gregory, W. K. and Milo Hellman

    1927 The Dentition of *Dryopithecus* and the Origin of Man. *Anthropology Papers of the American Museum of Natural History,* Part 1, New York.

Greulich, W. M. and S. Idell Pyle

    1959 Radiographic Atlas of Skeletal Development of the Hand and Wrist. Stanford University Press.

Helman, Milo
    1928 Racial Characters in Human Dentition. *Proceedings of the American Philosophical Society*, 2: 157-174.
Hertzberg, H. T. E.
    1968 The Conference on Standardization of Anthropometric Techniques and Terminology. *American Journal of Physical Anthropology*, 28: 1-16.
Hoerr, Normand L. and S. Idell Pyle
    1962 Radiographic Atlas of Skeletal Development of the Foot and Ankle. Charles C. Thomas, Springfield, Ill.
Howells, W. W.
    1969a Criteria for selection of Osteometric Dimensions. *American Journal of Physical Anthropology*, 30: 451-458.
    1969b The Use of Multivariate Techniques in the Study of Skeletal Populations. *American Journal of Physical Anthropology*, 31: 311-314.
    1970a Multivariate Analysis of Human Crania. Proceedings VIIIth International Congress of Anthropological and Ethnological Sciences, Tokyo and Kyoto. *Anthropology*, I: 1-13.
    1970b Mount Carmel Man: Morphological Relationships. Proceedings VIIIth International Congress of Anthropological and Ethnological Sciences, Tokyo and Kyoto. *Anthropology*, I: 269-272.
    1970c Multivariate Analysis for the Identification of Race from Crania. *In* Personal Identification in Mass Disasters. Edited by T. D. Stewart. National Museum of Natural History, Smithsonian Institution, Washington, pp 111-121.
Hrdlicka, Ales
    1907 Anatomy. Handbook of the American Indian, Part I. Bureau of American Ethnology Bulletin 30. Washington, D.C.
    1920 Shovel-shaped Incisors. *American Journal of Physical Anthropology*, 3: 429-465.
    1932 The Principal Dimensions, Absolute and Relative, of the Humerus in the White Race. *American Journal of Physical Anthropology*, 16: 431-450.
    1952 Practical Anthropometry. 4th Edition, edited by T. D. Stewart. The Wistar Institute of Anatomy and Biology, Philadelphia.
Hunt, Edward E. and I. Gleiser
    1955 The estimation of age and sex of preadolescent children from bone and teeth. American Journal of Physical Anthropology, Vol. 13, pp. 479-487.
Imrie, J. A. and G. M. Wyburn
    1958 Assessment of age, sex and height from immature human bones. British Medical Journal, Vol. 1, pp. 128-131.
Ingalls, N. W.
    1924 Studies on the Femur, General Characters of the Femur in White Males. *American Journal of Physical Anthropology*, 7: 207-255.
Jit, Indar and Shamer Singh
    1966 The Sexing of the Adult Clavicles. *Indian Journal of Medical Research*, 54: 1-21.

Johnston, Francis E.

1961 Sequence of Epiphyseal Union in a Prehistoric Kentucky Population from Indian Knoll. *Human Biology,* 23: 66-81.

1962 Growth of the Long Bones of Infants and Young Children at Indian Knoll. *American Journal of Physical Anthropology,* 20: 249-254.

1969 Approaches to the Study of Developmental Variability in Human Skeletal Populations. *American Journal of Physical Anthropology,* 31: 335-342.

Johnston, Francis E. and Charles E. Snow

1961 The Reassessment of the Age and Sex of the Indian Knoll Skeletal Population. *American Journal of Physical Anthropology,* 19: 237-244.

Keen, J. A.

1950 A Study of the Difference Between Male and Female Skulls. *American Journal of Physical Anthropology,* 8: 65-79.

Kelso, Jack and George Ewing

1962 Introduction to Physical Anthropology Laboratory Manual. Pruett Press, Inc., Boulder, Colorado.

Kerley, Ellis R.

1965 The Microscopic Determination of Age in Human Bone. *American Journal of Physical Anthropology,* 23: 149-164.

1970 Estimation of Skeletal Age: After About Age 30. *In* Personal Identification in Mass Disasters. Ed. by T. D. Stewart. National Museum of Natural History, Smithsonian Institution, Washington, pp 57-70.

Kraus, B. S.

1951 Carabelli's Anomaly of the Maxillary Molar Teeth. *American Journal of Human Genetics,* 3: 348.

1959 Occurrence of the Carabelli Trait in Southwest Ethnic Groups. American Journal of Physical Anthropology, 17: 117-123.

Kraus, B. S. and M. L. Furr

1953 Lower First Premolars. Part 1. A Definition and Classification of Discrete Morphologic Traits. *Journal of Dental Research,* 32

1953 Premolars. Part 1. A Definition and Classification of Discrete Morphologic Traits. *Journal of Dental Research,* 32: 554.

Krogman, W. M.

1962 The Human Skeleton in Forensic Medicine. Charles C. Thomas, Springfield, Ill.

Kronfeld, Rudolph

1935 Development and Calcification of the Human Deciduous and Permanent Dentition. *The Bur,* 35: 18.

Kummel, Bernard and David Raup, Editors

1965 Handbook of Paleontological Techniques. W. H. Freeman & Co., San Francisco, pg 852.

Lasker, G. W.

1950 Genetic Analysis of Racial Traits of the Teeth. *Cold Spring Harbor Symposium on Quantitative Biology,* Vol. XV. The Biological Laboratory, Cold Spring Harbor, Long Island, New York.

Lewis, Arthur B. and Stanley M. Garn
  1960 The Relationship Between Tooth Formation and Other Maturational Factors. *Angle Orthodontist,* 30: 70.
Lockhart, R. D., G. F. Hamilton and F. W. Fyfe
  1959 Anatomy of the Human Body. J. B. Lippincott Co., Philadelphia.
Ludwig, Fred J.
  1957 The Mandibular Second Premolars: Morphologic Variation and Inheritance. *Journal of Dental Research,* 36: 263-273.
Martin, Rudolf
  1928 Lehrbuch der Anthropologie. Jena (2nd ed., 3 Vol.).
McKern, Thomas W.
  1970 Estimation of Skeletal Age: From Puberty to About 30 Years of Age. *In* Personal Identification in Mass Disasters. Ed. by T. D. Stewart. National Museum of Natural History, Smithsonian Institution, Washington, pp 41-56.
McKern, Thomas W. and T. D. Stewart
  1957 Skeletal Age Changes in Young American Males, Analyzed from the Standpoint of Identification. Technical Report EP-45. Headquarters Quartermaster Research and Development Command, Natick, Mass.
Manouvrier, L.
  1893 La determination de la taille d'apres le grands os des membres. Mém. Soc. d'Anthrop., Paris, 2 ser., Vol. 4, pp. 347-402.
Middleton Shaw J. C.
  1928 Taurodont Teeth in South African Races. *Journal of Anatomy,* London, 62: 476-499.
Montagu, Ashley
  1960 An Introduction to Physical Anthropology. Charles C. Thomas, Springfield, Ill.
Moorrees, Conraad F. A.
  1957 The Aleut Dentition. Harvard University Press, Cambridge.
Morris, Henry and J. P. Schaeffer
  1953 Morris' Human Anatomy. Edited by J. P. Schaeffer. The Blakiston Co., New York.
Nelson, C. T.
  1937 The Teeth of the Indians of Pecas Pueblo. *American Journal of Physical Anthropology,* 23: 261-293.
Neumann, Georg
  1942 American Indian Crania with Low Vaults. Human Biology, 14: 178-191.
Oliver, Georges
  1969 Practical Anthropology. Charles C. Thomas, Springfield, Ill.
Ortner, Donald J.
  1968 Description and Classification of Degenerative Bone Changes in the Distal Joint Surfaces of the Humerus. *American Journal of Physical Anthropology,* 28: 139-156.

Pearson, Karl

1899 On the reconstruction of the stature of prehistoric races. Philos. Trans. Royal Society, London, Ser. A (Mathematical) Vol. 192, pp. 170-244.

1914-15 On the Problem of Sexing Osteometric Material. *Biometrika,* X: 479-487.

1917-19 A Study of the Long Bones of the English Skeleton I: The Femur. Dept. of Applied Statistics, University of London, Univ. College, *Drapers' Company Research Memoirs,* Biometric Series X, Chap. 1-4.

Pedersen, P. O.

1949 The East Greenland Eskimo Dentition. Meddelelser OM Gronland 142, Copenhagen.

Phenice, Terrell W.

1967 A Newly Developed Visual Method of Sexing the Os Pubis *American Journal of Physical Anthropology,* 30: 297-302.

Post, Richard H.

1969 Tear Duct Size Differences in Age, Sex and Race. *American Journal of Physical Anthropology,* 30: 85-88.

Powell, J. W.

1881 Annual Report of the Director. Bureau of American Ethnology Volume 1.

Pyle, S. Idell and N. L. Hoerr

1955 Radiographic Atlas of Skeletal Development of the Knee. Charles C. Thomas, Springfield, Ill.

1969 A Radiographic Standard of Reference for the Growing Knee. Charles C. Thomas, Springfield, Ill.

Redfield, Alden

1970 A New Aid to Aging Immature Skeletons: Development of the Occipital Bone. *American Journal of Physical Anthropology,* 33: 207-220.

Reynolds, Earl L.

1945 The bony pelvic girdle in early infancy. A roentgenometric study. American Journal of Physical Anthropology, Vol. 3, pp. 321-354.

1947 The bony pelvis in prepuberal childhood. American Journal of Physical Anthropology, Vol. 5, pp. 165-200.

Rippen, Bene van

1918 Mutilations and Decorations of Teeth Among Indians of North, Central and South America. *Journal of the Allied Dental Societies,* New York. September: 219-242.

Romero, Javier

1958 Mutilaciones Dentaires Prehispanicas de Mexico y America en General. Instituto Nacional de Antropologia e Historia. Mexico.

Schranz, D.

1959 Age Determination from the Internal Structure of the Humerus. *American Journal of Physical Anthropology,* 17: 273-277.

Schultz, A. H.

1930 The Skeleton of the Trunk and Limbs of Higher Primates. *Human Biology,* 2: 303-438.

1934 Inherited Reductions in the Dentition of Man. *Human Biology*, 6: 627-631.

Scott, J. H. and N. B. B. Symons
1958 Introductions to Dental Anatomy. Edinburgh.

Shaw, J. C. Middleton
1928 Taurodont Teeth in South African Races. *Journal of Anatomy* 62: 476-498.

Singh, Indera P. and M. K. Bhasin
1968 Anthropometry. Bharti Bhawan, Delhi, India.

Snow, Clyde C. and Earl D. Folk
1970 Statistical Assessment of Commingled Skeletal Remains. *American Journal of Physical Anthropology*, 32: 423-428.

Steele, D. Gentry and Thomas W. McKern
1969 A Method of Assessment of Maximum Long Bone Length and Living Stature from Fragmentary Long Bones. *American Journal of Physical Anthropology*, 31: 215-227.

Steele, D. Gentry
1970 Estimation of Stature from Fragments of Long Limb Bones. *In* Personal Identification in Mass Disasters. Ed. by T. D. Stewart. National Museum of Natural History, Smithsonian Institution, Washington, pp 85-97.

Stevenson, Paul H.
1924 Age Order of Epiphyseal Union in Man. *American Journal of Physical Anthropology*, 7: 53-93.
1929 On racial differences in stature long bone regression formulae, with special references to stature reconstruction formulae for the Chinese. Biometrika, Vol. 21, pp. 303-321.

Stewart, T. Dale
1940 Some historical implications of physical anthropology in North America. Smithsonian Institution Miscellaneous Collections, Vol. 100, pp. 15-50.
a Juan Comas en su 65 aniversario, Vol. II. Mexico.
1954 Evaluation of Evidence from the Skeleton. *In* R. B. H. Gradwohl, editor, *Legal Medicine*. C. V. Mosby Co. St. Louis. (See 1968 revision below.)
1958 The Rate of Development of Vertebral Osteoarthritis in American Whites and its Significance in Skeletal Age Identification. *The Leech*, 28: 144-151.
1962 Anterior Femoral Curvature: Its Utility for Race Identification. *Human. Biology*, 34: 49-62.
1963 New Developments in Evaluating Evidence from the Skeleton. *Journal of Dental Research*, 42: 264-273.
1965 The problem of analyzing the height of the cranial vault. In Homenaje
1968 Identification by the Skeletal Structure. *In* Gradwohl's *Legal Medicine*. Edited by F. E. Camps. The Williams and Wilkins Co., Baltimore.

Stewart, T. D. (Editor)

1970a Personal Identification in Mass Disasters. National Museum of Natural History, Smithsonian Institution, Washington.

1970b Identification of the Scars of Parturition in the Skeletal Remains of Females. *In* Personal Identification in Mass Disasters. National Museum of Natural History, Smithsonian Institution, Washington.

1970c Selected Bibliography on Personal Identification. *In* Personal Identification in Mass Disasters. National Museum of Natural History, Smithsonian Institution, Washington.

Stewart, T. D. and Mildred Trotter (Editors)

1954 Basic Reading on the Identification of Human Skeletons; Estimation of Age. Wenner-Gren Foundation for Anthropological Research, New York.

Telkkä, Annti

1950 On the prediction of human stature from long bones. Acta Anat. Basle, Vol. 9, pp. 103-117.

Thieme, F. P.

1957 Sex in Negro Skeletons. *Journal of Forensic Medicine,* 4: 72-81.

Todd, T. W.

1920 Age Changes in the Pubic Bone I: The Male White Pubic. *American Journal of Physical Anthropology,* 3: 285-334.

Trotter, Mildred

1934 Septal Apertures in the Humerus of American Whites and Negroes. *American Journal of Physical Anthropology,* 19: 213-227.

1970 Estimation of Stature from Intact Long Limb Bones. *In* Personal Identification in Mass Disasters. Ed. by T. D. Stewart. National Museum of Natural History, Smithsonian Institution, Washington, pp 71-83.

Trotter, Mildred and Goldine C. Gleser

1952 Estimation of Stature from Long Bones of American Whites and Negroes. *American Journal of Physical Anthropology,* 10: 463-514.

1958 A Re-evaluation of Estimation of Stature Based on Measurements of Stature Taken During Life and of Long Bones After Death. *American Journal of Physical Anthropology,* 16: 79-123.

Ubelaker, Douglas H., T. W. Phenice and William M. Bass

1969 Artificial Interproximal Grooving of the Teeth in American Indians. *American Journal of Physical Anthropology,* 30: 145-149.

Washburn, S. L.

1948 Sex Differences in the Pubic Bone. *American Journal of Physical Anthropology,* 6: 199-208.

1951 The New Physical Anthropology. *Transactions of the New York Academy of Sciences,* 13: 298-304.

Webb, C. H.

1944 Dental Abnormalities as Found in the American Indian. *American Journal of Orthodontics and Oral Surgery,* 9: 474-486.

Wetherington, Ronald K.

    1970 Laboratory Exercises in Physical Anthropology. Charles C. Thomas, Springfield, Ill.

Wilder, Harrison H.

    1920 Laboratory Manual of Anthropometry. P. Blakiston's Son and Co., Philadelphia.

In the Fall Quarter of 1972 I had 22 students in an upper division level class in Anthropology entitled Human Osteology of which six were graduate students. As additional course work for graduate credit the graduate students voted to index the text. The following index is a slightly revised version of their efforts supplied by:

FRANK N. CHARLES, III
BETTY ANDERSON FOSTER
BARBARA ANNE HALL
MARY JANE McCORMIC MULL
GUNNAR KENNETH MYKLAND
DIANA E. HUNT WITKOWSKY

To each and to all as a group I wish to express great appreciation for myself and for all of the future users of this laboratory and field manual.

William M. Bass
January, 1973

INDEX

177, Early Texas, 228, 236, Pima, 230, 231, 232, 233, 234, 236, 237, Pecos, 231, 232, 233, Pueblo Pecos, 234, 236, 237, Pueblos, 236, 237, Texas, 231, 232, 233

Indian Knoll, Kentucky, 115, 124, 131, 170, 187, 196, 234, 236, 237

Inferior, angle of scapula, 93, 94, articular processes of vertebrae, 79, 84, 85, basic orientation of body, 2, 244, border of ascending ramus, 47, border of horizontal ramus, 47, border of nasal, 44, border of scapula, 94, 103, border of squamosal of occipital, 35, 36, Iliac spine, 150, 151, nasal concha, 4, 30, 33, 42, 48, 49, 50, 51, 52, nuchal line, 37, 39

Infradentale, 56, 57, 59, 63, 72, superius, 56, 57, 59

Infra-orbital foramen, 42, 43, process, of scapula, 94, process, 45, 46

Ingalls, N. W., 168

Inion, craniometric point of skull, 57, 59

Innominate bones, 3, 5, 6, 34, 36, 39, 40, 88, 94, 148, 149, 151, 152, 153, 154, 155, 157, 159, 160, definition of, 243, plural of, 246, shape classification of, 242

Instep, 243

Instruments for anthropometric measurement, 9-11

Institute, Wistar Institute of Anatomy and Biology, 23

Institution, Smithsonian Institution, burial forms used by, 252

Intercondylar fossa, 164, 167

Intercondyloid eminence, 182, 184

Interlocking joints, or sutures, 33

Internal, basic orientation of body, 244, occipital crest, 39, occipital protuberance, 39, surface of rib, 107, 108

Internasal suture, 44, 59

International Congress of Anthropologists in Frankfort, Ger., 55

Interosseous, definition of, 243, crest, of fibula, 192, of radius, 3, 120, 122, 123, 129, 195, of ulna, 122, 126, 129, 195

Interpalatal suture, 60

Interrupted facet of the second cuneiform, 203

Intertrochanteric ridge, 176, 177, groove, 110, 111, 113

Introduction, 1

Irregular bones, 3, 7, 8, 135, 148, 150, cervical vert., 80, foot, 198, lumbar vert., 84, patella, 180, sacrum and coccyx, 86, thoracic vert., 82, vertebral column, 76, description of, 242

Irregularities in bones, 3

Ischial tuberosity, 150, 153

Ischiopubic ramus, 157, 158, 159

Ischium, 95, 102, 148, 150, 153, definition of, 243, plural of, 246

J

Japanese, 117, 228

Java, island of, 230

Jaw, 212, 213, 214, 228, lower, 29, 47, 61, 210, 213, 243, upper, 42, 210, 213, 243

Jit, Indar and Shamer Singh, 105

Johnston, Francis E., 115, 116, 124, 131, 170, 187, 196

Joint, ankle, 194, 195, for femur at hip, 160, hip, 166, sacroiliac, 148, 159, shoulder, 94, surfaces, 19, temporomandibular, 46

Journals, *of Forensic Medicine*, 21, *American Journal of Physical Anthropology*, 21, 22, 23, 97, *British Medical Journal*, 21

Jugular notch, 51, 90

Juncture, of crown root of teeth, 226

K

Kerley, Ellis R., 13

Kidney-shaped facet, of first cuneiform, 202

Knee, 166, 182, 184, 194, cap, 180, 181

Korean War Study, 13, 124

Kraus, B. S., 234

Krogman, W. M., 13, 16, 17, 21, 22, 97, 117, 156, 173

Kronfeld, Rudolph, 12, 14

L

L-shaped bones, 50, facet, of second cuneiform, 203

Labeling process for bones, 257, 258

Labial surface of teeth, 213, description of, 212

Labrador Eskimo, 228, 234

Lacrimal, 4, 30, 31, 33, 34, 42, 43, 48, 49, 51 crests of, 60, definition of, 243, duct of, 49, groove of, 43, 49, 50, plural of, 245, sac of, 49

Lacrimale, 56, 57, 60, 69

Lacrimo-maxillary suture, 60

Lambda, craniometric point, 57, 59

Lambdoidal border, of occipital, 37, suture of skull, 31, 33, 35, 36, 37, 38, 39, 59, definition of, 243

Landmarks of skull, 55, 56, 57, 58

Lapps, 234

Lateral, articular facet of patella, 180, basic orientation of body, 2, 3, 244, lateral border of nasal, 44, condyle of femur, 164, 172, 176, 181, 184, condyle of tibia, 182, 187,

Nasal, 4, 10, 30, 31, 33, 44-45, articulation of, with ethmoid, 48, frontal, 34, maxilla, 42, confusion of, with vomer, 51, aperture, 59, 60, 69, breadth, 61, 68, fossa, formation of, by maxilla, 42-43, formation of, by palate, 50, height, 61, 68, index, 62, 69, septum, formation of, by ethmoid, 48, formation of, by vomer, 51

N

Nasion, 56, 57, 59, 63, 67-68
Naso-frontal suture, 44
Naso-maxillary suture, 44-45
Nasospinale, 56, 57, 59, 68
Navicular, of the carpus, 135, 138, of the tarsus, 198, 199, 205
Neck, of the femur, 164, of the humerus, 110, 112, of the radius, 120, of the rib, 106, 107, of the tooth, 211, 212
Nose, 68, measurements of, 68-69
Nuchal crests, 73, lines, 37, 39
Nutrient foramen, of the femur, 113, 165, 167, fibula, 113, 195, humerus, 110, 112-113, metacarpals, 142, 143, metatarsals, 206, 207, phalanges, foot, 208, phalanges, hand, 147, radius, 113, 120, 122, tibia, 113, 182, 185-186, ulna, 113, 126, 128-129

O

Obturator foramen, 148, 151, 152, 162, groove, 3
Occipital, 4, 5, 31, 32, 33, 37, 38-39, 73, articulation of, with parietal, 35, sphenoid, 41, temporal, 39, condyles, 32, 37, 38-39, confusion of, with frontal, 34, parietal, 36, sphenoid, 42, temporal, 41, sutures, 33
Occlusal, 213, 224, 229, plane, 46, wear, 238-239
Occlusion, 239, 240
Olecranon, 126, 128
Olecranon fossa, 110, 115, 128
Opisthion, 58, 59
Opisthocranion, 57, 58, 59
Orale, 58, 59, 71
Orbital breadth, see measurements, skull, height, see measurements, skull, index, 62, 69
Orbitale, 56, 57, 60, 61, 67
Orbits, measurements of, 69, sexing of, 72
Orientation of the body, 1-3
Orthocrany, 64
Ossification, 14-16, of the carpus, 135, 136, 138, clavicle, 100, 101, femur, 164, 165, 166, fibula, 193, 194, humerus, 110-111, 112, 118, innominate, 148, 149, 150, 152, 155-156,

patella, 180-181, radius, 120, 121, 122, 125, ribs, 106, 107, sacrum, 87, scapula, 92-93, 97, tibia, 182, 183, 184, ulna, 126, 127, 128, vertebrae, 77-78
Osteoarthritis, 19, 62
Osteometric, board, 9, measurement of fibula, 187, equipment, 8-11
Osteometry, 54
Osteophytes, 19
Osteophytosis, 19, 20

P

Pacchionian bodies, 35, 36
Paired bones, 4, 5, 33, 75
Palatal breadth, see measurements, skull, index, 62, 71, length, see measurements, skull
Palate (palatine), 4, 5, 32, 33, 50-51, articulation of, with ethmoid, 48, inferior nasal concha, 51, maxilla, 42, sphenoid, 41, vomer, 51, measurements of, 70-71
Palatine process, 43, 50
Parietal, 4, 5, 30, 31, 33, 35-37, articulation of, with frontal, 34, occipital, 38, sphenoid, 41, temporal, 39, confusion of, with frontal, 34, occipital, 39, sphenoid, 42, temporal, 41, foramen, 35, 36, sutures, 33
Patella, 4, 5, 6, 8, 75, 180-181
Patellar articular surface, 164, 167
Pearls, 238
Pelvic girdle, 75
Pelvis, 148, 150, 151, sexing of, 156-157, 162
Permanent teeth, 13-14, 214-225
Perpendicular plate, 48, 49
Petrous portion, 39-40
Phalanges, 3, 15, 181, of the foot, 2, 4, 6, 19, 75, 198, 208-209, articulation of, with metatarsals, 206-207, difference of, from phalanges, hand, 147, proximal row, 208, 209, middle row, 208, 209, distal row, 208, 209, of the hand, 4, 6, 7, 75, 134, 146-147, articulation of, with metacarpals, 142, difference of, from phalanges, 147, ossification of, 135, 136, 138, proximal row, 146-147, middle row, 146-147, distal row, 146-147
Physiological length, ulna, 130
Pisiform, 135, 140, articulation of, with triquetral, 139
Planes, 1, 2
Platycnemic, 187
Platycnemic index, 187
Platymeric, 170
Platymeric index, 169-170
Platyrrhiny, 69
Ploughshare bone, vomer, 51
Plus 4 pattern, 211, 228, 229-230, 231, 232, 233

cervical, 80, thoracic, 82, lumbar, 84
Trapezoid, 135
Triquetral, 135, 139
Trochanters of femur, 16, 164
Trochlea of humerus, 110
Trotter, Mildred, 12, 22, 23, 24, 25, 26, 27, 117, 175, 197
Tubercle, 3, of rib, 106, greater multangular, 140, radius, 120
Tubular bones, 7
Turbulence, 52
Tympanic membrane, 52

U

Ubelaker, D., 241
Ulna, 1, 4, 6, 7, 16, 75, 111, 112, 122, 124, 126-132, 135, 138, 190, 195, 244, 246, age distribution stages of, 17
Ulnar notch of radius, 120, 122, 123, sides and directions, 244
Unpaired points, 55, 59

V

Vault of cranium, 7
Ventral arc, 157-159
Ventral, 2, 3, 244
Vertebra, 3, 4, 5, 6, 38, 53, 75, 76-89, 244, cervical, 80-81, thoracic, 82-83, lumbar, 84-85, sacrum and coccyx, 86-89

Vertebral column, 8, 76-89, osteoarthritis, 8, 9, 19, ribs, 107
Vertebrate body, 1
Vertebrochondral, 107
Vertebrosternal ribs, 107
Vertex, 57, 59, 244
Volar surface, 142-147
Vomer, 27, 32, 33, 42, 44, 48, 50, 51, 244, 245

W

Washburn, S. L., 54, 153-154, 156
Western Reserve Model Head Spanner, 11, 67
Western Reserve University, 12
Wilder, H. H., 131, 187
Wisdom tooth, 223-225
Wrist, 135-142

X

Xiphoid process, 90, 91, 244, 245

Z

Zygion, 56, 58, 60, 67
Zygomatic bone, 7, 30, 31, 32, 34, 39, 41, 42, 43, 45-46, 53, arch, 29, 30, 32, 39, 40, 45, 67, process, temporal bone, 40, 41, 74, maxilla, 43, 45
Zygomaticofacial foramen, 45

NOTES